Holographic Sub-Quantum Mind

Louis Malklaka

ISBN 1-508-81473-2

Table Of Contents

Acknowledgments

I would like to thank the scientists and researchers, past and present, living and non-living, upon whose achievements I have built. This includes, but is certainly not limited to such notable figures as James Clerk Maxwell, Michael Faraday, Nikola Tesla, David Bohm, and John Wheeler.

No amount of thanks could ever be enough to do justice to the accomplishments of such great scientific minds. They have provided the shoulders upon which our current generation of scientists stands. The only true and proper thanks is for us to provide a yet higher set of shoulders upon which *future* generations of scientists can one day stand. Only in this way can we hope to be worthy of what we have received, and to carry on the traditions of such esteemed predecessors.

I would also like to thank my "Beta Readers," especially William Sheppard, who went above and beyond, reviewing the book on a chapter-by-chapter basis as each chapter was completed, acting as my "target audience compass," helping to ensure that the book did not stray from its intended path. When he was able, for instance, to ask questions regarding the further implications of the material in the chapter he was reading; implications that, unbeknownst to him (as I had not yet provided him with a "table of contents"), were to be the subject of upcoming chapters; this let me know that he truly "got it," indicating that my target audience likely would as well.

And last but not least, I would like to thank my research assistant, Robin, without whose assistance with typesetting, editing, proof-reading, layout and design, as well as research, this book would never have come to fruition.

Introduction

This book, which began as a single volume, was cursed (or blessed, depending on how you look at it) by a severe case of *mission creep*. It was originally intended to consist only of the material in Part Two concerning the holographic sub-quantum nature of the mind. But, as so often happens in true scientific investigation—wherein you follow the facts *wherever* they may lead—the search for truth can sometimes lead you far afield of where you thought you were headed.

I initially assumed that *The Holographic Universe* by Michael Talbot—highly recommended as pre-requisite reading for this book—was going to provide enough of a foundation for the material in what is now Part Two. And in some respects, it still does. But when I began giving lectures on the subject of the holographic sub-quantum mind almost fifteen years ago, the questions from engineers and physicists in the audience made it very clear that the underlying physics of a holographic universe (and therefore, of the holographic sub-quantum mind) needed to be addressed. Research on that topic over the past fourteen years led to the creation of Part One of this book: a primer on sub-quantum physics.

The more I looked into the underlying physics of a holographic universe—in an attempt to verify what Talbot had written in *The Holographic Universe*—the more it led me to the conclusion that this book may have to become *two* books: the first being a primer on sub-quantum physics, with a working title of *Holographic Sub-Quantum Physics*. The second book, containing the material on the holographic sub-quantum nature of the mind, would naturally be titled *Holographic Sub-Quantum Mind*.

But then I realized that they really must remain as one book, because the research that now comprises Part One grew *out of* Part Two. So, while Part One now comes first—in order to provide the necessary infrastructure for the reader to comprehend the material in Part Two—

Part One probably never would have been written had I not set out initially to write what is now Part Two. Add to that the fact that terminology that is coined in Part One is referenced in Part Two, and you can see that, were someone to read the second book without having read the first, they would find themselves somewhat adrift.

Thus, what began as a single book evolved into two separate books that must, of necessity, remain conjoined twins.

The original book (now "Part Two") could have easily been written almost fifteen years ago. The newer half of the book (Part One) has consumed the intervening years.

Part Two deals with the eponymous subject: the holographic sub-quantum nature of the mind. This not only clears-up many mysteries concerning memory and cognition, and demonstrates that the mind resides, not within the physical brain, but *outside* of the body; it also provides a rational explanation for what have always been dubbed "paranormal phenomena." By the end of Part Two, it will be quite clear that such phenomena are not paranormal at all; it is simply that, until now, the laws of physics have not been able to explain how such events are possible.

In this book, there are certain theories that are taken as *givens*, as I consider the authors of the referenced books to have done outstanding work in supporting their hypotheses. I highly recommend that you follow-up, and read the referenced books for yourself. They are not a pre-requisite—the reader can simply take their premises as givens in order to read *this* book—but I highly recommend that you do your due diligence by reading the referenced books.

At the top of the list, I highly recommend *The Holographic Universe* by Michael Talbot. He does such a spectacular job of pulling together the evidence, it really is a *must read* for anyone interested in the holographic nature of the universe. Talbot's book shows that the universe *is* holographic; the first chapter of *Holographic Sub-Quantum Mind* shows *why* it is holographic.

Other recommended reading (in no particular order):

- Any papers by David Bohm. (His work is more technical than Talbot's book; but if you are able, it is worth reading. If not, Talbot does a *very* nice job of summarizing Bohm's work in his book; so do not feel that you are "missing out" at all.)
- *The Big Bang Never Happened* by Eric Lerner
- *The Electric Universe* by Wallace Thornhill & David Talbot (no relation to Michael Talbot)
- *The Trouble With Physics* by Lee Smolin

The theories to be taken as givens based upon the aforementioned works:

1.) The universe is holographic in nature (Talbot & Bohm)
2.) The universe is infinite, physically and chronologically (Lerner & Thornhill/Talbot; and by extension, Hannes Alfvén)
3.) That gravity plays a much smaller role in the movement and structure of the cosmos than has heretofore been supposed (Lerner & Thornhill/Talbot; and by extension, Hannes Alfvén)
4.) That electricity plays a much *greater* role in the movement and structure of the cosmos than has heretofore been supposed (Lerner & Thornhill/Talbot; and by extension, Hannes Alfvén)
5.) That String Theory has missed the mark, and the Standard Model is in need of extensive revision in order to make it viable going forward (Lee Smolin)

Please note that, while I am taking the works of these authors as givens, this is not intended to indicate that they support (or are even *aware of*) my work; merely that I am building upon their material: i.e., that they have put together a body of work that is compelling enough for me to consider their points as having been made, and for me to therefore accept their assertions as givens for the purpose of this book.

Any mistakes that I make going forward are mine, and mine alone, and should not reflect negatively upon the work of these authors.

Complementary, Not Contradictory

This book does not *contradict* Bohm, nor even (for the most part) standard Quantum Mechanics—it is *complementary* to them.

Bohm and others have within their theories a "black box" labeled "somehow"—it is *within* that box that we find the sub-quantum. Bohm's "Implicate Order" (his theory's "black box") *is* the sub-quantum.

The so-called "fundamental particles" (which are actually neither) are nothing but the result of interference patterns between sub-quantum entities. This explains *why* the universe is holographic. That is why this book *complements* Talbot's work: whereas he shows that the universe *is* holographic in nature via numerous examples, Part One of this book shows *why* it is holographic; why it cannot be otherwise. This is known as the "fractal principle"—anything built from smaller components will ultimately resemble those smaller components, if you look closely enough. A universe built upon interference patterns will, inevitably, appear holographic, once we learn to recognize what we are seeing.

As Lee Smolin so correctly pointed-out in his book, *The Trouble With Physics*, there may be a lot *wrong* with modern Physics (his main example: *String Theory*), but there is also a lot that is *right*. Never throw the baby out with the bath water. Keep what is right, and correct what is wrong. That is how science progresses.

Take Newton, for example: his laws may not work well at the quantum level, but they work amazingly well in day-to-day situations. Newton understood the "what"—movements of and interactions between physical objects that he could actually measure, and create formulas to describe with mathematical language—but not the *why*.

Quantum Mechanics is similar: theorists can see the *what,* but they concoct ridiculous explanations such as "The particle *knows* when you're looking at it!" They use this foolish notion to explain the results of the double-slit experiment, because they do not know the *why*. Once they do, the ridiculous theories will fade away, and be replaced by reality. But the *observations* (the "what") actually *work* (much like Newton's laws at the day-to-day level) for engineering devices and systems; so we will keep what works, and simply update our understanding of the *why;* which will eventually improve our ability to engineer devices and systems.

Thus, we will not *scrap* Quantum Mechanics; we will simply make it *better.* Keep what works, and update what needs updating. Again, that is how science progresses.

Too many fringe-science "nut-jobs" (at least, that is how they are inevitably labeled by the scientific Mainstream) think they need to start over from scratch, and formulate an entirely "new science" with their own terminology, rather than working within the existing framework to the greatest extent possible; and—in most cases, rightly so—their work is resoundingly ignored or ridiculed by the Mainstream. If they had only resisted throwing out the baby with the bath water, at least a few of them could (potentially) have made positive contributions to the advancement of science.

That is why it is always best to take the logical approach: keep what works, and fix only what *needs* fixing. Or, as the old adage has it, "If it ain't broke, don't fix it!"

Many people ask (usually during the Q&A sessions at the end on my seminars) "Hey, if we 'don't understand how the universe works,' then how can we make particle accelerators, TVs, computers, and all kinds of other highly advanced technology?" They fail to comprehend the difference between science and engineering. You can engineer devices and systems without understanding the *deepest-level* underlying physics. As long as you can measure the *what,* the ultimate *why* is optional.

That said, a better understanding of the *why* can definitely facilitate more efficient devices. For instance, one day, instead of strapping humans to giant explosives (rockets) in order to travel into space, we will use anti-gravity systems that are built by people who finally understand how gravity works, and who then design devices and systems to actually *control* it rather than fighting *against* it.

One should not assume, simply because engineers can design systems based upon their understanding of the *what,* that this indicates they necessarily understand the *why.* That would be equivalent to assuming that, just because someone is an excellent driver, they are also an expert auto mechanic who understands exactly how everything under the hood functions. Many of Hollywood's best stunt drivers would probably be surprised if people were to make that assumption about them.

This illogical assumption is similar to the logical fallacy "Post hoc, ergo propter hoc"—not *exactly* the same, but similar.

Another example (since I have already mentioned anti-gravity)—ask ten different physicists how gravity works: they will either have ten different theories, or they will say "I don't know." Perhaps both. But they can engineer devices and systems just by measuring *what* gravity does, and create formulas to *describe* (math is a *language*, remember) what it is doing. You do not need to understand the *why* in order to engineer devices and systems. Much the same as with driving a car: you do not need to know *how* it works (i.e., internal combustion specifics, electricity and spark gaps, hydraulics, torque converters, etc.) in order to *drive* it. Nor do you need to be an electronics expert to turn on and use a TV or computer.

Another example is an auto mechanic. (Going the other way for a moment, since we already used expert drivers as an example.) A good auto mechanic may be able to diagnose a bad alternator, and replace that alternator. They may even have a vague idea of how an alternator works. Not *what* it does—they all understand that—rather, *how* it actually functions from a Physics standpoint. It would be the rare auto mechanic who actually understands all of the "nitty-gritty" Physics/ Electrical Engineering involved in how an electrical generator specifically works. Not that I am conceding that modern Physics/Electrical Engineering truly understands the actual deepest-level *why* of it either. But for the purpose of illustration, you take my meaning. We will deal with the sub-quantum basis of current flow and magnetics—not to mention gravity—later in the book.

Once again, being adept at working with devices and systems—even to the extent of being able to diagnose and repair them—does not require (or provide *proof of*) an expert knowledge of *why* they work.

Another example (the last one, I promise): doctors. A "General Practitioner" (GP) may be able to diagnose a heart problem; but he is not a "Heart Specialist." He knows enough to recognize that there may be "something" amiss with your heart; so what does he do? He sends you to a *specialist*. That GP is probably the *last* person you want standing over your spread-open chest trying to perform a quadruple bypass!

But getting back to the main point: there is no need to *negate* the whole of the Standard Model—only to *complete* it. (No need to re-invent the wheel.) Yes, we may negate portions of it—the portions (such as String Theory) that make assumptions/propose hypotheses regarding the deepest levels of reality—but by-and-large, we will keep the *what*—which we know works rather well—and simply work on the *why*. This can only improve our ability to engineer devices and systems as our understanding of the *why* improves.

People who use Quantum Mechanics on a daily basis to engineer devices and systems do not concern themselves with such ridiculous hypotheses as "The particle *knows!*" Such silliness does not impact their day-to-day work of engineering working devices and systems. They only need to know the *what* for that.

When we document a new fact regarding the *what,* engineers go to work immediately utilizing that effect to either improve existing devices and systems, or to create entirely new devices and systems that could not have been created before that new fact was discovered. But while the engineers are going to work using the new fact, the theorists get to "work" spinning (often ridiculous) hypotheses about the *why* of it all. They can spin beautiful, self-consistent mathematical fictions for years-on-end (while absorbing copious quantities of government grant money) to justify ludicrous, illogical notions about the *why* of it all; while the engineers simply put the newly discovered fact to actual use.

Of course, the general public assumes that, since the engineers are busily creating new widgets (or improving old ones) as a result of the new discovery, that necessarily means that the ridiculous hypotheses being put forward by the math-spinning theorists actually hold water. Further confusing the issue is the aspect of the math itself. When challenged, theoretical physicists often protest "But the math works-out!" We then have to remember that mathematics is simply a *language*. And like any language, it can be used to tell the truth, or to tell lies. Just because the math is self-consistent and checks out, does not make the science built upon that math necessarily true. Science cannot be built strictly upon a house of cards made of math. If your initial assumptions are incorrect, no amount of "beautiful math" in-between your assumptions and your erroneous conclusions will save you.

Rational scientists deserving of the appellation simply follow the facts to where they lead, rather than trying to shoe-horn them into some pre-conceived notion. That is what science is supposed to be about.

A good example of such true science in action is the recent NASA "Comet Impact" experiment. NASA had a theory about what *should* happen per their "comets are dirty snowballs" hypothesis. The Electric Universe people (aka "Plasma Cosmologists") made quite a different prediction concerning the outcome of the experiment, based upon the Electric Universe model.

When the 800 lb. copper projectile impacted the comet, "NASA Scientists Perplexed" is how the headlines read. They had no idea *why* what had happened, happened. But—as you may have guessed—what happened is *exactly* what the Electric Universe crowd *predicted* would happen.

That is the truest test of any science: the ability to predict outcomes. Which is exactly why String Theory has not been verified by a single experiment in all its decades of existence. It is time for logic and common sense to return to science. What follows is my small contribution to that effort.

Part One:
Sub-Quantum Physics

"All truths are easy to understand once they are discovered; the point is to discover them." — Galileo Galilei

Louis Malklaka

Chapter 1:
The Double-Slit Experiment

[The double-slit experiment] "has in it the heart of quantum mechanics. In reality, it contains the only mystery."
                                            — Richard Feynman

It all began with that accursed double-slit experiment. The real shame of it all, is that it held such *promise*, if only the results had been interpreted correctly.

If someone had only realized what the experiment revealed: that, at the most fundamental level of reality, the universe is holographic. What we see around us is an emergent phenomenon arising from the fractal expression of the interactions between fundamental interference patterns taking place below the "inviolable barrier" known as the Planck Length.[†]

The improper interpretation of the results of this experiment steered physics down a dead-end road, from which it has not recovered to this very day. It has transformed what used to be the most strictly rational of all the sciences into a field now dominated largely by "magical thinking."

You may already be familiar with the specifics of the double-slit experiment, and all of the permutations thereof that have taken place over the years. If not, you may think that when I present my "in a nutshell" explanation below, that I am either having a joke at your expense, or greatly exaggerating. But I assure you, this is exactly what is actually being taught in modern Physics classes all over the world. I encourage you to verify this for yourself by looking it up in any standard Physics textbook, via Google, or even on YouTube—there are many videos that will explain the double-slit experiment in accordance with the brief synopsis I provide below. As crazy as it may sound, this is what is being taught as "science" these days.

The double-slit experiment demonstrated that light behaves like waves in water. If you have ever seen ripples on the surface of water interact with each other, you are familiar with what an interference pattern looks like. (Fig. 1) If you put a piece of some

Fig. 1

---

† In the Standard Model, the Planck Length is the theoretical minimum size that anything can be.

solid substrate in the way of a single wave, and cut two slits in the substrate, two new waves will be emitted on the other side of the substrate; one from each slit. These two new wavefronts will then interfere with each other in a picture-perfect textbook fashion, as they are each identical to the other, having been created at the same time, from the same initial wave, by two identical slits in the substrate. When we shine a monochromatic light source at a double-slit device, we do not see the actual interference pattern in the air as we do with water, since light must shine *upon* something for it to be seen; but we can see the resulting interference pattern on a target screen. (Fig. 2)

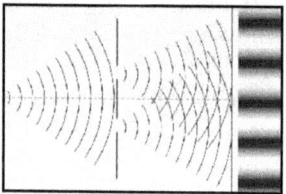
Fig. 2

This result led physicists to conclude that light behaves as a wave. So far, so good. But when experimenters decided to "watch" the slits to get a closer look at what was happening, the interference pattern disappeared. The image on the screen looked similar (though still slightly more diffused: more on that later) to what one might expect if a series of BBs were launched at the target; i.e., the light seemed to behave as if comprised of "particles"[†] when experimenters were watching.

Here is where a sane, logical person would conclude that, by "watching" the slits to see what came through, they were somehow *interacting* with the process, and therefore changing its behavior. Unfortunately, this was not the conclusion reached by mainstream physics in this case. They determined that the particles of light could "somehow" act as waves under certain circumstances—such as when they were forced through a double-slit arrangement—but would act as particles when we observe them; as if light was actively hiding its secrets

---

[†] I enclose the word "particles" in quotes here, because not only is light *not* a "particle;" there is actually *no such thing* as a particle. Most modern physicists actually realize this, but continue to refer to the most fundamental units of matter that we have thus far identified as "particles," both for ease of explanation, as well as calculation. And since all new particle physics students are told to "Shut up and calculate!" when they ask troubling questions, we can see that math is placed *far* ahead of truth in modern mainstream physics. Much more on this in later chapters.

from us. This is known as the "Wave-Particle Duality" problem of modern physics.

Aristotle taught that contradictions cannot exist, and that when we believe we have encountered one— as we seem to have done in the case of the double-slit experiment—we must examine our premises, as one of them is certain to be faulty. In this case, the faulty assumption made by mainstream physics is that the so-called "fundamental particles" (which are actually neither. Again, more on this later) are the most fundamental building blocks of our physical reality.

When we realize that there is another level (if not a plethora of levels) *below* the hallowed Planck Length; and that the interference patterns produced at this *sub-quantum* level are responsible for guiding the movements of all "particles" that exist *above* the Planck Length, the results of the double-slit experiment now make perfect sense. After all, in our water example above, what would happen if you were to block one of the slits in the substrate? The interference pattern would disappear.

An example of this "wave guiding a particle" effect can be seen in the wave tank experiments of French physicist Yves Couder. His work was replicated by Daniel M. Harris and John W. M. Bush from MIT's Department of Mathematics. You can watch the video here: http://youtu.be/YF5iHQMjcsM

Thus, we can see how what we will, for the moment, call "particles" (for lack of a better term; until chapter three) can produce an interference pattern, even when only single particles are fired one-at-a-time at the double-slit[†] device. Yes, this actually happens. And, as you may have come to expect, mainstream physics has yet another ludicrous explanation for this: that the particle somehow "interferes with itself." There is a lot of math attached to that statement, using (or, more specifically,

---

† Even when only a *single* slit is used, we still see some diffusion in the pattern that appears on the target screen; exactly as we would expect, since the sub-quantum elements that are flowing freely through the single slit will "fan out," much like water. The standard explanation (again, convoluted and "reaching") is that the diffusion is due to "particles deflecting off of the sides of the slit." Once again, Occam's Razor to the rescue: the simplest explanation is usually the correct one, especially when it consistently explains multiple phenomena. That is always to be preferred over separate convoluted explanations for the same phenomenon in different situations.

*mis*using) the Schrödinger Wave Equation; but in essence, that is what all of the math is saying. We must remember that math is only a *language,* and like *any* language, it can be used to tell the truth, or to tell *lies*.

When we realize that an interference pattern at the underlying sub-quantum level is *guiding* the particles, there is no longer a contradiction, and no longer a need for bizarre, convoluted explanations. The interference pattern we can actually *see* is simply a higher-level fractal expression† of the *sub-quantum* interference pattern. Occam's Razor is fully served.

All of the incorrect theoretical physics that has arisen over the past eighty-plus years due to improper analysis of the double-slit experiment has been a house of cards built upon faulty assumptions. We may be getting ahead of ourselves a bit here, but you can probably imagine how eliminating the contradictions and incorrect assumptions about the most fundamental nature of our physical reality might just lead to a simple, common explanation for electricity, magnetism, and gravity. But as I said, we are getting ahead of ourselves a bit: we will discuss that further in later chapters.

In the meantime, we must ask: "So, what exactly *is* this 'sub-quantum' level of reality?" Put simply, it is what Nikola Tesla called the *dynamic ether*. (As distinguished from the incorrect notion of a *static* ether.) But whereas Tesla (and most of his contemporaries) simply took the existence of an ether for granted and looked no further; in Chapter 2, we will delve more deeply into the specifics of the sub-quantum, picking-up where Tesla's generation left off.

Oh, I know, I know; I can already *feel* the screams of protestation. It is forbidden by the scientific mainstream even to *breathe* the word 'ether.' Doing so will quickly get one exiled to the 'lunatic fringe' of pseudo-science, and thoroughly ostracized by the greater scientific community. The only way to make such a major blasphemy any *worse*, is by mentioning the name "Tesla" in that same breath.

---

† Similar to the "Golden Ratio," aka the "Golden Section," or "Phi" with regard to life. It is encoded into the DNA of all living things, so we see this ratio expressed it in all life forms on Earth.

So, we will ignore Tesla. Forget I even mentioned him. After all, he wasn't *really* a physicist, but merely an electrical engineer; so what could *he* possibly know, right? (Forgetting for a moment that Newton was an alchemist, Faraday was a bookbinder and valet, and Einstein was a patent clerk. But still . . .)

In the interests of not ruffling *too* many mainstream feathers (or has that ship already sailed?), let us by all means move on to a much more *respectable* source squarely from the mainstream of physics, in the personage of David Bohm.

Bohm's take on the double-slit experiment was closest to the truth. He believed that a particle is guided through space via a wave he called the "quantum potential." So while a "particle" may only physically go through *one* slit, the quantum potential wave goes through *both*, thereby creating the interference pattern that guides the particle to impact only in certain areas on the target screen. Bohm further believed that this "quantum potential" existed *below* the Planck Length, at a *sub-quantum* level, and that this quantum potential explained the observations and measurements of quantum physics.

As you can see, he was *so close*, missing the full truth only *slightly,* by A.) not specifying what this "quantum potential" wave actually is, and B.) insisting on the "particle" idea. But as Isaac Newton once famously said "If I have seen further than others, it is by standing upon the shoulders of giants." And as I said in the introduction, sub-quantum physics is *complementary* to the ideas of Bohm and others— even to Quantum Mechanics—not *contradictory.* And in this case, it can be said that Bohm took us most of the way there; he did the "heavy lifting," as it were, to such an extent that his *quantum potential* can be considered to be synonymous with what I term the *sub-quantum vortices.* (More on that in Chapter 2.) I begin where Bohm left-off, simply adding detail to his sketch, as modern technology has allowed us insights not available in his day. Had such technology been available to him, I am quite certain I would not be writing this book, as he would have beaten me to it by quite a number of decades.

One final bit of information on Bohm comes via a quote from John Stewart Bell (author of *Bell's Theorem*): ". . . In 1952 I saw Bohm's paper. His idea was to complete quantum mechanics by saying there are certain variables in addition to those which everybody knew about. That impressed me very much."

And as if the opinions of David Bohm and John Stewart Bell were not sufficient to show that we are on firm footing (which is precisely why David Talbot spent so much time on Bohm in *The Holographic Universe*), let us turn our attention to none other than John Wheeler.

Toward the end of his career, Wheeler arrived at the conclusion that the "zoo of particles" of the *Standard Model* were simply a higher-level manifestation of a deeper, simpler level of reality. He felt that the mechanics of this simpler level, once surmised, would restore rationality to quantum mechanics, obviating the need for the mystical explanations of which the "pop science" culture has become so enamored. Wheeler further believed that an understanding of this "deeper, simpler level" would eliminate the "fuzziness"[†] we find at the boundary between the quantum and classical worlds.

Wheeler wrote the following poem to express his ideas as discussed above:

Behind it all
is surely an idea so simple,
so beautiful,
so compelling that when —
in a decade, a century,
or a millennium —
we grasp it,
we will all say to each other,
how could it have been otherwise?
How could we have been so stupid
for so long?

---

† Just as cutting a holographic film into hundreds of pieces still allows you to display the entire image from the smallest piece; but the image gets "fuzzy" as the pieces get smaller due to the limitations of the wavelength of the laser being used (the laser that *must* be used, since it was used to create the hologram in the first place); the "fuzziness" we see when looking into the quantum world is a result of trying to use "pieces of the hologram" to study its finer layers. Perhaps we simply need finer tools.

Chapter 2:
The Tripartite Universe

Pythagoras was more correct than he knew when he stated that the universe was comprised of three parts.

Certainly, much like the four cardinal directions of the compass can be further divided into finer degrees, there are undoubtedly finer divisions in-between the three main levels of reality; but just as the finer gradations of the compass points do not invalidate the four cardinal directions, neither does the inevitable eventual discovery of additional degrees of fineness in-between the three main levels of the tripartite universe invalidate those three main levels.

The first of the three levels is the one with which we are most familiar: macroscopic reality—everything down to the so-called "fundamental particles" (which are actually neither) and the "inviolable" barrier known as the Planck Length.

The second level was discussed in chapter one—the "ether," or sub-quantum vortices: the finer infrastructure *below* the Planck Length which is responsible for the odd behavior of the Double-Slit Experiment.

The third and final level (to which I will refer hereafter as the "pixel grid"—more on that in a moment) is the smallest "foundation level" of reality†—pure angular momentum, or a "tendency to spin."

Why "pixel grid?" Because it helps to think of it as a three-dimensional version of the pixel grid on any

---

† We may one day discover that even *finer* components make-up even the infinitesimal entities of this level. Just as with the so-called "Cardinal Directions" of North, South, East and West, where there are finer degrees of divisibility (from NE, SW, etc., to even *more* specific/finer delineations) between those cardinal points; so, too, we understand that each of these layers specified above is within itself near infinitely divisible. But just as the cardinal directions are still useful (e.g., simply say "Go East") without giving some finer level of specificity; it is useful to simply say "sub-quantum vortices" (or *ether*) and "sub-quantum 'pixel grid.'"

Another example is a rainbow. We can see seven basic colors in a rainbow. And yet, there are (on average) up to ten *million* colors that are discernible to the human eye. Those seven colors in a rainbow are the broadest color break-down; but there are *millions* of colors *in-between* those seven.

In much the same way, when we say above that there are "three levels" to reality, we are aware that, similar to the rainbow, there may be *many* levels of refinement *in-between* those three basic levels, and that our scientific instruments may one day be able to detect at least a *few* more of them. In the meantime, just as knowing that light can be split into seven "basic" colors with a prism can be very useful (and *engineerable*), we can advance scientific knowledge of the universe's most fundamental infrastructure considerably with the knowledge of these three basic levels.

We have come quite far in identifying and quantifying the macroscopic level—everything above the Planck Length—but the other two levels are currently "Terra Incognita;" so simply knowing they are there, and beginning to investigate them will bear enough fruit to keep scientists and engineers busy for many decades to come.

The Tripartite Universe

1.) The "Sub-Quantum Pixel Grid"—comprised of infinitesimal "tendencies to spin," or "pure angular momentum." This most primary level of physical space is incompressible. Thus, it is the "pixel grid" upon which the rest of our holographic universe is "displayed." Its incompressible nature is what allows instantaneous transmission of data to every point in the universe simultaneously.

2.) The "Sub-Quantum Vortices"—secondary vorticular structures produced by interference effects (holographic) between the individual units of the sub-quantum pixel grid. This level of the sub-quantum (i.e., below-Planck-Length) physical world is an intermediary between the sub-quantum pixel grid and the so-called "fundamental particles" of the Standard Model.

3.) The "Fundamental Particles" (e.g., electrons, quarks, etc.) are, in reality, tertiary vorticular constructs, comprised of conglomerations of sub-quantum vortices.

television screen or computer monitor. To aid in this visualization, picture the sub-quantum pixel grid as a jar full of marbles, packed-in so tightly that there is no possibility of movement. Now, think of the jar itself as a "plane" in geometry: you have to draw edges on the plane for the textbook, so they tell you to simply visualize that the plane goes on forever. The same is true with the jar and its contents: picture the marbles going on forever, effectively eliminating the jar, since an infinite universe cannot be contained.

The individual elements of the infinite sub-quantum pixel grid, being incompressible and packed-in every bit as tightly as our marbles in the example above, cannot move. But each one can rotate within its own confines; similar to a person "running in-place," going nowhere fast. But where our little "pixels" differ is that they are vortices; so while they cannot "move" *per se*, they *can* convey information to their neighbors. In this way, they actually act more as tiny gears than marbles. But whereas physical gears would "lock-up" if placed together cheek-by-jowl in this manner, the sub-quantum pixel grid elements, are non-physical (i.e., not macroscopic, as is the world with which we are intimately familiar); and thus, no lock-up occurs.

We can see how whatever happens to any single one of these pixel grid elements is instantly felt by *all* of the rest of them in the infinite universe. This is how data is stored (holographically, I might add) in the infinite sub-quantum infrastructure. It is *information* that is conveyed from each unit to the next, not physical motion. Thus, rather than the "lock-up" we would see when *physical* gears meet in opposition; instead, what is produced in the sub-quantum infrastructure is an *interference pattern*. "Ahhhh," I can hear the holographically-astute reader exclaiming; for interference patterns are at the very heart of how holograms are produced. And thus, while Michael Talbot and others have more than adequately proved that the universe *is* holographic in nature; now we know *why* this is so. This is the very definition of holographic: all of the data

Relative Sizes[*] Of Sub-Quantum Elements And "Fundamental Particles"

Smallest "Fundamental Particles" of the Standard Model (Quarks, etc.)

Sub-Quantum Vortices / "Ether"

Sub-Quantum "Pixel Grid"

[*] Note that this is a *two*-dimensional analogical representation, and is not intended to be to-scale.

for the whole is stored within each individual piece. Every pixel grid element is like David Bohm's "drop of ink in glycerine" that has been "stirred," literally *forever*, thereby "enfolding" (to use Bohm's term) all data about every interaction it has ever experienced. This makes the sub-quantum pixel grid, not merely a "holographic projection" system (more on that in a moment), but an infinite universal holographic storage medium.

When I say "projection," it is an *analogy only*. There is no actual movement, just as modern mainstream physics shows that there is no actual "matter." (Any physicist worth his salt will tell you that "solid matter" is, in reality, nothing but *lots* of empty space, and energy fields.)

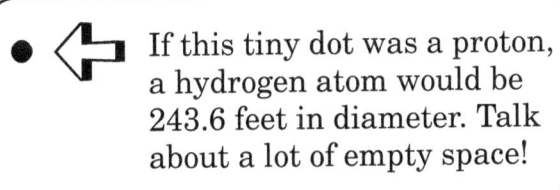

If this tiny dot was a proton, a hydrogen atom would be 243.6 feet in diameter. Talk about a lot of empty space!

So, what is actually meant when we say that the sub-quantum pixel grid is a "holographic projection" system? To grasp the answer, you must first understand that the universe is holographic. (Remember the "givens" mentioned in the introduction?) Not only did Michael Talbot do an excellent job of pulling-together research across several decades showing this to be the case, even modern "mainstream" physics is now accepting that the universe is holographic; though many of them have a misunderstanding of the mechanics involved. (For example, renowned Stanford Physicist Leonard Susskind. He and others believe the holographic universe is "projected" from a two-dimensional membrane surrounding our "curved space-time." This is an incorrect conclusion resulting from their acceptance of Einstein's flawed premises.) Nonetheless, they *do*, at least, agree that the universe is holographic. They could hardly do otherwise, as the evidence is overwhelming. Thus, proceeding from this basic knowledge that the universe is a hologram, we can then realize that nothing we perceive as "solid" or "real" is anything of the kind, but merely a "convenient illusion" that allows us to interact with each other and the world around us. And since holograms are nothing but "interference patterns," we can immediately see how interference patterns generated between the sub-quantum pixel grid elements produce the infinite hologram that is our universe. It is really quite simple when you think about it; but then, basic truths usually are.

This brings us to the next level up from the pixel grid: the sub-quantum vortices; what Tesla called the "dynamic ether[†]." These sub-quantum vortices are the first-level holographic constructs projected by the pixel grid. As such, while the pixel grid is static, the sub-quantum vortices are free to "move," i.e., *appear* to move. If you think of the pixel grid as the "movie projector" and the sub-quantum vortices as the "film," you can see how this illusion of movement occurs. After all, the movies you watch on TV or at the theater are nothing more than an *illusion* of movement created by displaying a specific number of frames-per-second (usually 30 fps for TV, and 24 fps at the movies). If you look at a reel of movie film, you can see each individual frame, one-at-a-time, and the illusion of motion disappears. Similarly, while the sub-quantum vortices *appear* to "move;" they are, in fact, simply being "displayed" in different locations as circumstances dictate. And since macroscopic objects (the "fundamental particles" of the standard model, and everything comprised thereof) are simply collections of sub-quantum vortices, we can see that everything from the sub-quantum vortices upward is simply being "projected" by the sub-quantum pixel grid[††].

---

† Many critics of the very concept of an "ether" object that such a medium would have to be *incredibly* dense, and near "infinitely rigid." We have *exactly* that in the sub-quantum pixel grid.

†† The "fundamental particles" are simply the interference patterns between groups of sub-quantum vortices. It is similar to whirlpools in a river: they are not really a solid, separate "entity," merely a movement of the water. The difference is, we can *see* the water. Imagine if we couldn't see the water, but *only* the whirlpools. Those "fundamental particles" are the "whirlpools" in the infinite sea of interfering sub-quantum vortices. Thus, the so-called "fundamental particles" are, in actuality, tertiary vorticular constructs.

This is, in fact, what is responsible for the "speed of light" limitation, as every projection system has a "refresh rate"—the maximum speed at which a system can cease projecting one image, and begin projecting another. Every TV, computer monitor, cell phone screen, etc., has a refresh rate. The higher that rate, the smoother the video. The refresh rate of the sub-quantum pixel grid is such that, in most circumstances, macroscopic matter (meaning, in this context, everything from the Planck Length upward) is limited to "moving" at the speed of light. Entities *below* the Planck Length are not limited by the speed of light, as the less you are breaking-down and re-building, the faster it can be accomplished. Yes, there will undoubtedly still be an upper limit to the refresh rate, but the higher the complexity of the object being projected, the longer it will take; and thus, the lower the maximum speed at which that object can appear to move. The *less* complex an object, the faster it can appear to move; all the way down to information, which is conveyed instantly. Sub-quantum communication would thus be like the proverbial "tug on a rope"—information would be conveyed instantly across infinite distances. (If that sounds suspiciously like Tesla's "scalar wave communications" system, then not only do you "get it," you have really done your Tesla research.)

What about sound? It is not a physical "thing," so why does it move so much more slowly than light? Because with sound, it is not that you are conveying a "particle†;" you are conveying the specific state of vibration of a *collection* of particles: a much more *complex* set of parameters. Such a complicated system is going to take much longer to break-down and re-build than something as (relatively) simple as light.

Having shown thus far that there is no matter and no motion, it is easy to see how time (at least the "linear time" that we experience) is similarly illusory. Just as

---

† Note that, while I may use the term "particle" for convenience, even mainstream modern physics understands that what are generally referred to as "particles" are, in reality, "vorticular energetic structures." Because it is more convenient to refer to them as "particles," for simplicity's sake, I will adopt that convention for the remainder of the book.

running a movie in "fast forward" mode (with speeds from "2x" to "32x" being available on many modern set-top video players) allows us to alter the rate at which images are displayed (or frozen with the "pause" feature; or even slowed-down with "frame-by-frame" mode), the concept of "absolute time" in the macroscopic universe is a fiction. A movie or TV show is designed to be played at "1x," so that people can comprehend what is occurring; i.e., the "time scale" of the video matches the day-to-day experience of the viewers. Similarly, while events that unfold in "physical reality" generally take place at "1x," that is simply the "standard mode" of the holographic projection that we call the "physical world." It is entirely possible that we may one day develop the technology with which to engage "2x" (or higher) modes; or even "pause" or "reverse." (This would make "time travel" a real possibility. But more on that in the second half of the book, as this diverts us into such topics as the "many worlds" interpretation of quantum mechanics.)

Speaking of quantum mechanics, I wish to reiterate here that the sub-quantum levels as discussed in no way *contradict* the majority of the tenets of quantum mechanics. Rather, they *complement* them; and in reality, *complete* them.

One of the aspects of quantum mechanics with which *many* prominent physics have taken issue (including Einstein. Whatever his other errors, he was right to object here) is what Einstein called "spooky action at a distance." Considering the instantaneous information transfer of the sub-quantum pixel grid, "action at a distance" need no longer be considered "spooky," as we can (eventually) identify *exactly* what causes each and every reaction we see. Yes, it will require the creation "from scratch" of a new instrumentality capable of measuring sub-quantum elements; but that is nothing new. We had to do exactly that when electricity and magnetism were first discovered, and our scientific and technological infrastructure of the day proved equal to the task. And again when radiation was discovered. Modern science is *much* more prepared to accomplish this feat than was our rudimentary science of several hundred years ago. By comparison, developing sub-quantum measurement tools and systems should be a "piece of cake." It simply requires first seeing that there is a *need* (i.e., that the sub-quantum actually *exists*, as people like Tesla, Bohm, Wheeler, et al have been saying for over a century); then the need can be filled. Necessity truly *is* the mother of invention.

The addition of sub-quantum physics to quantum mechanics can eliminate all of the objectionable "spooky weirdness," because now that we have identified the mechanism, standard Newtonian "cause-and-effect" rules can be applied. Since *probability* currently reigns supreme at the quantum level, that is our first clue that something important is missing. Given proper data, we will one day be able to predict quantum events as accurately as we predict macroscopic events.

The failings of quantum mechanics (e.g. "The particle *knows* when you're looking at it!" explanation for the results of the double-slit experiment) have simply been due to not realizing that a finer infrastructure exists below the Planck Length. (Despite the assertions of the previously-named scientists.) Thus, generations of scientists and engineers have not been *looking* for anything at that level, and theoretical physicists have instead spun webs of mathematical mumbo-jumbo (such as string theory) to account for what, to them, seemed "spooky." Now that we have ample evidence that something *does*, in fact, exist below the Planck Length, scientists and engineers can finally correct the failings of the theoretical physicists, replacing their fantastical, mystical notions with cold, hard science; as it always goes. Once again, science can replace myth with reality. From "thunder is made by the gods" to understanding what is, in reality, a "sonic boom." We can finally dispense with such myths as string theory with its multiple 'hidden dimensions,' and replace it with something that reflects what we actually find below the Planck Length.

So again, sub-quantum physics is not at all a *replacement* of quantum mechanics; simply a minor alteration: the scientific equivalent of going to the barber to get a trim. But in our case, we will be "adding a little to the bottom" rather than "taking a little off the top."

Backing-up a bit now to the sub-quantum pixel grid elements themselves; how do we know they are "vortices?" Quite simply, because—unlike string theorists—we follow the evidence.

In chapter one, we showed that the foundation level of reality is sub-quantum; i.e., below the Plank Length. But with regard to the actual *nature* of that sub-quantum level; what does the evidence indicate? Extrapolating from all of our observations at the macroscopic level, where vorticular motion is everywhere; we can see that, as we get smaller, things look even *more* vorticular, (angular momentum/ spin at the sub-atomic level, etc.), indicating that whatever exists below the Planck Length *must* be vorticular in nature. All more complex systems are explainable in terms of their simpler infrastructure. This is known as *fractal expression*[†]. When you peer down into the deepest level of reality, the picture becomes *simpler*, not more complex. At the base-most level, we literally have nothing at all; nothing more than pure angular momentum: a "tendency to spin."

The chain of evidence leading us inexorably to the discovery of a vorticular substructure to the universe began with Anaxagoras (circa 500 BCE),

† Fractal expression; or as the Freemasons say, "As Above, So Below." When Hermes Trismegistus penned this axiom, did he realize the full scope of his statement? To put fractal expression in the simplest terms possible (i.e., without all of the heady math utilized in typical "fractal analysis"): anything made-up of smaller pieces will exhibit certain characteristics of those pieces. If you build something from Legos, it is going to look "blocky," unless you stand *very* far back. But look too closely, and you will see the inherent "blockiness." When we simply go about our daily lives and observe macroscopic "matter" with our macroscopic eyes, it is similar to standing extremely far away from a Lego sculpture: if well-built, it will look "smooth." But if you observe the sculpture from close-up, the illusion of smoothness quickly fades. Modern scientific investigations have allowed us to look more closely at the sub-structures of matter, and the "blockiness" (i.e., the true vorticular nature of reality) has revealed itself. Everywhere we look, from galaxies to the sub-atomic level, we see spin/rotation/angular momentum. If we understand fractal expression, this clearly indicates that whatever lies *below* what we have identified thus far must also be rotational/vorticular in nature. This is why we see "vorticular" elements even at the macroscopic level (19.5 degrees north and south latitude "hot spots" on rotating planetary bodies, hexagonal cloud formations at the north poles of planets, etc.)—fractal expression, from the very base layer of the sub-quantum pixel grid, to the sub-quantum vortices, to our "fundamental particles" with their electron clouds/shells, all the way up to macroscopic matter.

who was the first to conceive of this possibility. Had Maxwell's equations not been translated out of their original quaternion math into standard algebra, thereby resulting in various situations of *vorticular* stress all being lumped together as "zero resultant vector" situations, thus hiding whatever might be going on internally in these situations; the vorticular nature of our universe might have been proven much sooner.

Unfortunately, it would not be until the 1940s that Soviet physicist Lev Landau[†] would be the first modern physicist to revisit this idea. By the 1970s, Dr. Winston Bostick[††] and Dr. Daniel Wells[†††] were working together to develop a fluid vortex model of the electron.

More recently, in an extended series of experiments at the University of Michigan, Alan Krisch and his team showed that protons are deflected as a result of collision more frequently when their spins are parallel rather than in opposition. Additionally, the protons deflect to the left almost three times as often as to the right. In other words, the protons behave as exactly as we would expect if they were tiny *vortices*.

It is important in science to always follow the data where it leads, regardless of any preconceived notions you may have. Unfortunately, this is something that the founders of string theory failed to do. It is a shame that string theory came relatively close to the truth, missing the mark only with regard to size scale (they were afraid to go below the inviolable Planck Length barrier), and unwisely choosing "strings" instead of *vortices*—especially odd, considering that all of the math points to vortices, and string theorists

---

[†] An expert in fluid mechanics and quantum field theory who received the 1962 Nobel Prize in Physics.

[††] From 1956 to 1981 he was the George Meade Bonde Professor of Physics at Stevens Institute of Technology in Hoboken, NJ (department head until 1968), and researched plasma vortex phenomena. He was also a consultant to the Los Alamos National Laboratory and the Lawrence Livermore National Laboratory.

[†††] Dr. Wells worked at Princeton Plasma Physics Laboratories in the 1960s, and was a former student of Dr. Bostick.

are so enamored of math†. Instead of hypothesizing "strings" at random, the founders of string theory should have followed the *evidence*, which clearly points to *angular momentum*, which requires *rotational motion*. A "vibrating string" cannot produce angular momentum. This should have let them know from Day One that "strings" were not the answer. Granting, for a moment, the benefit of the doubt; perhaps the founders of string theory chose strings, because they remembered from their earliest Physics lessons the example of "Mary and Sally are holding a jump rope and using it to make waves." Forgetting, of course, that vortices *also* make waves. (Just observe a whirlpool in any body of water: it radiates waves outward in every direction.) But of these two phenomena—both of which can generate *waves*—which one can generate angular momentum? Yes; only a *vortex*.

---

† Quantum Physics equations abound with h (Planck's Constant) divided by 2π. (For example, an electron's orbital angular momentum is given by $h/2\pi$, $2(h/2\pi)$, $3(h/2\pi)$, etc.) For this reason, physicists created a new placeholder for $h/2\pi$: $\hbar$, pronounced "*aitch bar*." Since we know that a circle's circumference is given by $2\pi r$, it is quite revealing to see h involved with 2π so frequently. Again we see fractal expression, demonstrating the relationship to the sub-quantum *vortices*.
Further, engineers researching energy at NASA's Lewis Research Center discovered that the equations describing the fundamentals of energy are identical to the equations for vorticular fluid flow.

Chapter 3:
Waves vs. Particles

As we saw in the first two chapters, the entire universe is a holographic construct; so none of it is "real" in the way that we think of objects as "solid." There is no such thing as a "particle." While *we* may perceive what seem to be "particles" at *our* level of perception, it is all simply an illusion; a "projection" of the sub-quantum pixel grid; just like watching a show on TV. None of it is real—no actual motion: simply dots on a screen.

Everything is built from the interference patterns created when sub-quantum pixel grid elements interact. So, in reality, the answer to the question: "Waves or Particles?" should really be "neither waves *nor* particles: it is all *vortices*."

Think for a moment of a video game: it is simply a computer program, with images being displayed on a screen; an illusion. None of what happens in the game is "real." But within the parameters of the computer program that controls the video game, there are rules that must be obeyed. Your character cannot, for instance, simply run through a wall: s/he has to go through doorways. The wall in the game is not "real;" merely an artificial construct. But you have to operate within the rules of the program, which dictate that all of the artificial constructs interact with each other in a certain predefined fashion.

In much the same way, the universe is merely a holographic projection; but all of the holographic constructs interact with each other in a specific way. So, within that context, it still behooves us to have the discussion: "Waves vs. Particles."

The reason that there is no such thing as "wave-particle duality," is that there *are* no "particles," just as there is no such thing as "solid matter." (Any mainstream physicist worth his salt will readily admit this, and explain it to you at length.)

But those same physicists who, on the one hand, will explain to you that nothing is "solid"—that it is all simply force fields pushing against each other, giving the *illusion* of solidity; like trying to push two strong magnets together (and they are 100% correct in this assertion)—

will turn right around and say that the "Photo-Electric Effect" (among other things) proves that there are "solid particles." They cannot have it both ways. Did they forget that *nothing* is actually "solid?" The scientific mainstream conveniently *forgets* this fact when it comes to the nonsensical "wave-particle duality" madness. The Standard Model's fictitious "particles" are no more solid than the rest of the matter in the universe: just another case of force fields pushing against each other.

The so-called "particles" that are regularly "bounced" off of each other in accelerator experiments are, in reality, no more solid than your hand and the wall; and they "bounce" off of each other for the *same reason* that you cannot push your hand through the wall: force fields in repulsion. Just because the *scale* changes, does not mean the "particles" are any more solid than our illusory macroscopic "solid matter." Physicists seem to forget this, simply because the "particles" are so small.[†] They still seem intent upon finding "solid objects" at some base level.

Once we dismiss the delusion of wave-particle duality, we can more easily grasp the true sub-quantum infrastructure of reality.

As we saw in chapter one, it is an *underlying interference pattern* that guides the "particles" in the Double-Slit experiment to impact the target in such a way as to display an interference pattern; a representation of the basic underlying infrastructure that lies below the Planck Length.

Since interference patterns are a *wave phenomenon*, anything that arises therefrom must *also* be a wave phenomenon. (There is that *fractal expression* again.) We can see that the conglomeration of sub-quantum vortices commonly referred to as a "particle" is actually an *emergent phenomenon* arising from *interference patterns*. Thus, a "particle" is, in actuality, a "wave packet"—i.e., a bundle of waves.

---

† λ (wavelength) = h (Planck's Constant) / p (momentum) — Meaning that wavelength and momentum are inversely proportional. i.e., Increasing a particle's momentum *decreases* its wavelength. Thus, as "wave packets" are accelerated to greater speeds, they may *seem* more "point-like" due to the decrease in wavelength, but they are still "wave packets."

Since I have mentioned the "guiding" of the "particles" in the Double-Slit experiment by waves, it behooves us to discuss for a moment the physicist who first discovered this effect.

Louis de Broglie explained in his 1924 PhD thesis (which won him the Nobel Prize in 1929) why only certain "orbits" were allowed in an atom: the electrons' wave patterns would destructively interfere with themselves after one orbit unless the orbital size was an even multiple of the electron's wavelength.[†]

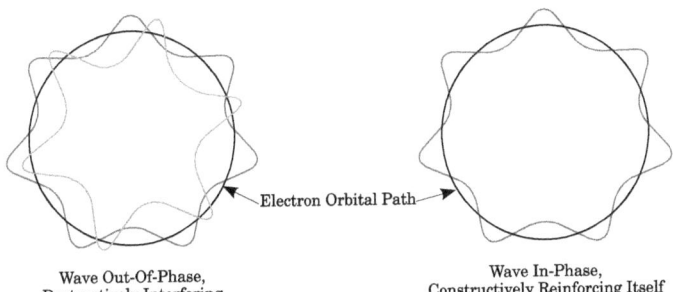

Wave Out-Of-Phase,
Destructively Interfering
With Itself

Electron Orbital Path

Wave In-Phase,
Constructively Reinforcing Itself

Applying this model to the electron orbits of the hydrogen atom, de Broglie noticed a perfect agreement between his wave model of the electron and the experimentally determined energies of a hydrogen atom. The lowest-energy state (aka 'ground state') has one wavelength encircling the atom. The next energy level (aka the 'first excited state') has two wavelengths encircling the atom. The third excited state has three wavelengths encircling the atom. This pattern continues flawlessly for each succeeding energy level.

So we can readily see that, while de Broglie's idea of an electron as a "particle" that was *accompanied* by a wave that acted as its "carrier" was wrong, his basic idea is correct: i.e., that the "wave packet" called an "electron" has a specific wavelength, and only *even multiples* of that wavelength can exist within an atom.

† It should be noted that, whereas de Broglie's theory only considered orbit length, in reality, radius must also be considered, since the electron's "wave packet" encompasses the entire 3-dimensional space within the atom.

Add-in different "phases" to accommodate the "Pauli Exclusion Principle,"† and we have a decent picture of the atom.

What are generally called "particles" are actually "wave packets"—conglomerations of sub-quantum vortices of unique, characteristic configurations that we identify as individual "particles," just as specific waveforms are identified as "words" in voice-recognition software. And much like the wave packets used in voice recognition, as we build-up a "vocabulary" of these wave packets, we end-up with an absolute *zoo* of so-called "particles," just as we see within the "Standard Model" of Physics.

Just as a language has an alphabet that makes-up individual words; so, too, there is an "alphabet" for the wave packets that comprise our fundamental "particles." And just as an alphabet is much simpler than the zoo of wave packets that accumulates as we identify different *words*, the "alphabet" of discrete combinations of sub-quantum vortices would be much easier to identify and understand than the veritable *dictionary* of wave packets that comprise the Standard Model. Yes, you need to know "words" in addition to an alphabet; but knowing words without understanding the alphabet would make you like a tourist in a foreign country who only memorizes certain words and phrases rather than actually learning the language. To truly "learn the language" of our universe, we have to learn the *alphabet* of the sub-quantum vortices.

If I record an audio file of the word "Hello," you can look at the waveform—merely a visual representation of a bundle of vibrational patterns—and say "That represents the word 'Hello.' " That "wave packet" we see on the computer screen represents a specific "word." But that does not change the fact that what we have is simply a discrete packet of vibrations. We may *interpret* it as a "word;" but that does not change what it actually *is* at its most fundamental level: simply a packet of vibrations.

---

†The *"Pauli Exclusion Principle"*—only electron "wave packets" that are *out of phase* with each other can co-exist within an atom, since they can inter-penetrate without interference.

Similarly, referring to the various discrete "fundamental" entities that comprise the Standard Model as "particles" does not change the fact that they are *not* "particles." They are, rather, discrete groupings of wave forms that we recognize as various specific entities; just as a specific "wave packet" of vibrations we may recognize as the word "Hello" does not magically transform that bundle of vibrations into anything other than what it is.[†] "A rose by any other name," as Shakespeare would remind us.

From an organizational standpoint, referring to the various wave packets of the Standard Model as "particles" may be *useful*—much as utilizing a compiled dictionary of the different shapes of "wave packets" for voice recognition is useful—but it does not change the fact that referring to something as a "particle" does not make it so. The problem is exacerbated when "bosons" come to be thought of as "real particles" (there's an oxymoron for you); when it is well-known that a "boson" is a "fake" particle; meaning an effect that is *treated* (mathematically speaking) like a particle, but that scientists know is *not* actually a particle. But mathematically (as well as conceptually), it is *convenient* to treat it as a particle. Unfortunately, this leads to confusion; especially when even scientists who should know better start to think of bosons as "real particles." This happens all too often when scientists and mathematicians mistake *representations* of reality for reality itself. An old Zuni saying comes to mind: "Do not confuse the map with the terrain."

This is what Einstein failed to understand when he wrote his 1905 paper on the "Photo-Electric Effect." (For which he received an undeserved Nobel Prize in 1921.) Considering the sound analogy above as an example: the waveform that represents a specific word may be a

---

† Imagine someone speaking to you in Chinese. If this stream of sound was recorded digitally, it would still produce a group of wave packets that represent discrete words; but unless you understand Chinese, those sounds are *meaningless* to you. And if your computer does not "understand" Chinese (i.e., has not been programmed for Chinese voice recognition/translation), they are just meaningless wave packets to *it* as well. The truth of the matter is revealed: those "words" have no actual meaning; they are simply discrete packets of vibration. The meaning is *relative*, not *intrinsic*. Similarly, a wave packet commonly known as some type of "particle" may produce a specific *effect*—just as an *understood* word will—but that does not make it a solid "particle."

discrete packet of sound (as opposed to a long, ongoing waveform, as with recorded music); but that does not mean that it is not still a sound. Our innate tendency to mentally categorize and separate things has allowed the wool to be pulled over our eyes, trying to make us believe that "A" is somehow *not* "A." But it *is!* A sound is still a sound, and a wave is still a wave, regardless of whether it is a discrete bundle, or comes in the more typical continuous form; even if it comes in a discrete packet that sounds like the word "Hello." It is still a *wave.* Just as what are termed "particles" are, in reality, simply *wave packets*. Thus, the entire "Wave vs. Particle" argument is moot, and the notion of "Wave-Particle Duality" a complete fiction.

In 1927, researchers Clinton Davisson and Lester Germer of Bell Labs in New Jersey discovered that, when electrons impacted a crystalline surface, diffraction and interference effects were produced. By measuring the predominant angle at which electrons reflected from the surface of the crystal, combined with the previously determined internal atomic geometry of the crystals, Davisson and Germer were able to calculate the wavelength of the electrons utilized in the experiment.

It should be noted that diffraction and interference are *wave phenomena*. This is further evidence that the electron is a "wave packet," and *not* a "particle."

We can see how the quantization Einstein attributed to "particles" in his paper on the "Photo-Electric Effect" can easily be explained via the concepts of *resonance* and *harmonics*. Einstein incorrectly concluded that light was acting like a particle due to the requirement that the light be of a specific *frequency* rather than have a certain *intensity* in order to complete the task of dislodging electrons from the impacted surface. His faulty logic: if light was comprised of particles, more *intensity* simply meant more of the *same sized/energy* "photons" (as the light "particles" were called) with the energy level of *each photon* being determined by its frequency.

Unfortunately, Einstein had it all wrong. Since the electron is a wave-based entity, the reason that a specific *frequency* of light is required is that the frequency of the applied light must match (or be a close harmonic of) the frequency of the electron in order to achieve sufficient *resonance* to move the electron. We do not need "particles" to see that "wave packets" will all resonate at the fundamental frequency of a system as well as its *harmonics.*†

So, now that we have put the "wave vs. particle" question to bed once and for all; in the following chapters, we can move on to discuss the many *applications* of sub-quantum physics.

---

†This is why striking any "E" key on a piano will cause the "E" strings on a nearby guitar to resonate.

Chapter 4:
Gravity

"It is inconceivable that inanimate brute matter should, without the mediation of something else which is not material, operate upon and affect other matter, without mutual contact, as it must do if gravitation in the sense of Epicurus be essential and inherent in it. And this is one reason why I desired you would not ascribe 'innate gravity' to me. That gravity should be innate, inherent, and essential to matter, so that one body may act upon another at a distance, through a vacuum, without the mediation of anything else, by and through which their action and force may be conveyed from one to another, is to me so great an absurdity, that I believe no man who has in philosophical matters a competent faculty of thinking can ever fall into it. Gravity must be caused by an agent acting constantly according to certain laws; but whether this agent be material or immaterial, I have left to the consideration of my readers."

—Isaac Newton, in his third letter to Bentley, 1692

While the first three chapters were necessary to lay the foundation for the Holographic Sub-Quantum Mind, this chapter and the remainder of "Part One" may not seem strictly "necessary"—at least, not on the surface—as applied *directly* to the concept of the Holographic Sub-Quantum Mind. But for the sake of completeness, as we move forward with Gravity, Magnetism, Electricity, etc., try to realize that, in a way, they *are* necessary; because if you can understand how all of the basic elements of Physics arise from the sub-quantum infrastructure (as Bohm, Wheeler, Tesla, et al believed), it will be easier in the second half of the book to grasp how consciousness is simply "one more phenomenon" that *also* emerges from the same sub-quantum infrastructure.

The subjects that comprise the remainder of "Part One" are (at least to some) interesting enough in their own right, and could easily have comprised their own book, as I mentioned in the introduction. But they are also key to comprehending the *second* half of the book; not only by having established a foundation, but because any time you have a "one-off" (signal vs. noise—signals repeat, while noise cancels), the tendency is to say: "Eh; maybe, maybe not." But when you see a *vast* number of phenomena that are *all* explained by the same underlying Physics (and not just "explained by," but explained in a more coherent manner than the multitude of hackneyed, cobbled-together, ad hoc theories strung-together by mainstream Physics to explain the same phenomena—Occam's Razor, and all that); it makes it easier to see how consciousness (and other phenomena generally lumped-together as "paranormal," though they are nothing of the kind) can spring from the same solid foundation.

As you drill down into the finer levels of any system, things should become *simpler*, not more complex. So, by covering all of the Physics aspects in this *first* half of the book, the *second* half is seen for what it truly is: a natural extension of all that was shown in "Part One." Consciousness becomes an "of course."

Gravity is the result of a sub-quantum pressure differential. In effect, gravity is the result of a universal pressure that "pushes" on everything; and when matter gets in the way of that pressure, it creates a pressure differential that results in a net acceleration in the direction of the mass that blocked said pressure.[†]

Picture a giant beach ball (say 6-feet in diameter) in the middle of a large room, such as a gymnasium. Place "hurricane fans" around the edge of that room, in a circle, surrounding the giant beach ball, and turn them all on simultaneously. The beach ball will remain in the center of the room. Now, move a large object (a piano, refrigerator, etc.) into the circle: it will block the fans from that direction, allowing the *other* fans to blow the beach ball *into* the large object. This is how a sub-quantum pressure differential produces what we call gravity: as the sub-quantum vortices flow through objects, the objects (temporarily, like light traveling through water) slow the sub-quantum vortices down (a processes known as sub-quantum "drag"), thereby allowing the sub-quantum pressure from all *other* directions to press the objects together.

In the beach ball example above, you can *feel* the wind produced by the hurricane fans; so it is obvious to you what has happened. But imagine if you *could not* feel the wind: you might assume that the large object somehow "attracted" the beach ball to itself by virtue of its mass. This is what leads to the illusion that objects "attract" other objects. As you sit there reading this, the reason you do not simply float away is *not* that the Earth is "pulling you down;" it is that the sub-quantum pressure that had to travel through the Earth is pushing *up* on you less strongly than the rest of the universe is pushing *down* upon you.

Since gravity is the result of a sub-quantum pressure

[†] Stars are round (and likely planets—since they are largely liquid before solidifying; deformation of solid lithosphere bodies must take place later, after they have become solid) due to sub-quantum pressure pressing in on them from all sides. We can see this behavior (see video at http://youtu.be/erC0ykitS6s) in water spheres on the International Space Station. The extent to which *stars* are out of round could tell us something about how the sub-quantum pressure reacts to electrical discharge. We already know from electro-gravitics that there *is* such a reaction.

differential, this tells us that there must be a specific *maximum* level for gravity. Different densities of matter may shield the sub-quantum pressure to varying extents; but once you hit 100% shielding, no additional pressure differential can be produced; and thus, no stronger "gravity" will be possible. The same is true for magnetism and electricity, as we will see in the next two chapters.

Think of a sail boat: once you have a sail that blocks 100% of the incoming wind, no additional thickness or impermeability of sail material is going to add to your speed. The speed of the wind is the speed of the wind: period. Once you have a sail that can capture 100% of that energy, no further thickness of sail will help you. If you started with chicken wire, and worked your way *up* to, say, canvas; *then*, yes, you would see an improvement with each improvement in sail material, as you would be blocking (capturing) an increasing percentage of the incoming wind. But once you hit 100%, the improvement stops: you are at the maximum possible.

The same holds true for gravity: once we have a material (or thickness, EM field, etc.) that blocks 100% of incoming sub-quantum pressure, there will be no additional increase possible in the acceleration of gravity at the surface of the object in question (planet, moon, asteroid, or even a star), as the sub-quantum pressure is the sub-quantum pressure: period. Just like the wind in a ship's sails.

Einstein, Acceleration, And Inertia

Einstein's famous "Equivalency Principle" equated gravity and acceleration. But what does that really *mean,* at a deeper level?

When we realize that what we call "gravity" is produced by a sub-quantum pressure differential, it becomes all too clear that, of *course* gravity and acceleration *must* be equivalent; or at the very least, closely related.

What the Standard Model refers to as the "Higgs Ocean" is *analogous* to the sub-quantum plenum, though they still fear to break the "Planck Length" barrier. But it has often been pondered, Why does the so-called "Higgs Ocean" only resist *accelerated* motion?

Understanding as we do now the sub-quantum pressure differential explanation for gravity, answering this age-old question is actually the simplest thing of all: since the sub-quantum pressure is the same from all directions, when an object is either at rest, or moving at a steady rate of speed, the pressure from the front that is resisting forward motion is being countered by the exact same amount of force from behind. But to *accelerate* the object to a *higher* rate of speed would require that *additional* force be applied from behind in order to over-balance the opposing force from the front. This is known as "inertia."

Quantum Gravity

We can see now how the sub-quantum pressure differential explanation for gravity leads to solving the "Quantum Gravity Problem" in the Standard Model of Physics. If we apply our sailboat analogy to a boat paddle, we can quickly realize why there is essentially no gravity at the quantum level.

If we create a thin (but sturdy, so it is strong, yet does not provide much resistance when swept through the water) metal *frame* of a boat paddle, then cover it with chicken wire; we can see that this will not be very useful for rowing a boat, as it glides through the water with virtually no resistance; no "drag." Replace the chicken wire with "hardware cloth" (square holes much smaller than chicken wire), and while you may be able to measure the increase in resistance with a sensitive force gauge; still not much use for rowing. Move up to window screen material, and you may begin to notice more usefulness for rowing, as the reduced aperture size of the holes in the screen provides more resistance when the paddle is moved through the water. Finally, we move to something more solid—such as canvas—and now we can generate some useful force when rowing.

Similarly, at the quantum level, there is insufficient substance to produce enough "drag" to create any appreciable resistance to the flow of the sub-quantum vortices. Thus, there is essentially "no gravity" at the quantum level. The agglomerations of sub-quantum vortices that we call "fundamental particles" are simply not large enough (when compared to the sub-quantum vortices) to produce enough of a sub-quantum pressure differential. Only once we reach larger *groupings* of sub-quantum vortices (aka "macroscopic matter") do we finally have enough "drag" to create a sub-quantum pressure differential sufficient to produce what we call "gravity."

The Illusion Of Density

A question I am asked quite often is: "What about density? If you hold out your empty hand, that's quite different from your hand with a gold bar in it. How can it be that the empty hand and the hand with the gold bar in it are the same? How can the 'stuff' that's there when your hand is empty be the same 'stuff' that's there when your hand is full of something solid?"

The problem with this question, is that the querent is limiting themselves to *our* perspective; i.e., a *macroscopic* perspective. A television set would not make much sense to a caveman, as he would have no frame of reference. It would be counter-intuitive from his "every day" perspective.

Give that same caveman two über-strong class 52 rare earth magnets (held in opposition; i.e., N-N or S-S; otherwise, *ouch!*), and he will not be able to comprehend why he cannot force them together, since there is nothing but "empty space" separating them. It would be "magic" to him.

It "feels" as though there is something solid in-between the magnets keeping them apart; but that feeling of "solidity" is an *illusion*. There is nothing but a "force field" in between those magnets. Similarly, modern mainstream physics has known for many decades that "solid matter" is no such thing: it is mostly *empty space*. Thus, "empty hand" vs. "gold bar in-hand" is really the same from an absolute perspective. The apparent "density" of the gold bar is as much an illusion as the "magic force" keeping those two strong magnets apart. What you "feel" as a "gold bar" in your hand is nothing more than one set of illusory holographic projections interacting with another. And just as your television does not care which program you watch (it really does not matter, as they are *all* illusions. None of those shows or movies is actually "moving." That is an *illusion* of movement created for you on the screen. The pixels are there regardless of what content you choose to make them display), the sub-quantum pixel grid does not care whether it "projects" a hand full of "empty air," or a gold bar; because in either case, the number of "pixels" used does not change. The density of the sub-quantum "pixel grid" and sub-quantum vortices remains the same regardless of what they are "displaying." The difference between an "empty hand" and a "hand full of gold bar" is as illusory as anything you might choose to display on your TV. There *is no difference.* (Or, to be more specific; the only difference—just as with which TV show you choose to display—is the *information.*)

Since the holographic projections we call "matter"—including the "fundamental particles"—are in reality not "solid" at *all*. It is merely the interactions between these illusory "fundamental particles" and their similarly illusory holographic sub-components (the sub-quantum vortices) that give us the emergent illusion known as "density;" in much the same way that two powerful rare earth magnets in repulsion mode exert a great amount of force against each other, while two non-magnetized pieces of metal of the exact same composition, dimensions—and therefore, density—exhibit no repulsive force at all. The alignment of the sub-components of the magnets are responsible for the repulsion effect, as it affects how they internally channel the sub-quantum pressure.

This also explains the amazing difference in strength between the electrical force and the gravitational force (the electrical force is 10^{39} times as powerful as gravity): the (relatively) weak gravitational force—which is produced via a "drag" effect of the sub-quantum vortices—is easily overpowered by the *much* stronger electrical forces between the holographic projections we call "matter." This illustrates the difference between a "passive" force such as gravity (i.e., produced merely due to temporarily slowing-down the sub-quantum vortices as they pass through matter), and the more powerful forces of magnetism and electricity that result from substances which *actively* channel the sub-quantum pressure, to much greater effect.

And as far as *actual* density goes; the sub-quantum pixel grid is the "densest" medium you could hope to imagine. (Our smallest "fundamental particles" are comprised of massive agglomerations of units—sub-quantum vortices—that are *holographically projected* by the even *smaller* base elements of the pixel grid.) So *actual* density is no problem. What we *perceive* macroscopically as the "density" of an object is as illusory as the field between the above-mentioned magnets; or as matter itself. And just as a minor re-arrangement of the atoms inside two "chunks of metal" transforms them into extremely powerful magnets; a very slight change in the arrangement of the sub-quantum vortices in your "empty hand" can transform that "empty space" into a gold bar; because in an absolute sense, there truly *is no difference.* (More on this in the "Manifesting" sub-section of "Chapter 22: Magic(k).")

Anti-Gravity?

So, understanding now how gravity actually works; let us have a bit of fun by engaging in some speculation. How might someone like Ed Leedskalnin have built his "Coral Castle" in Florida?

Since gravity is the result of sub-quantum vortices experiencing "drag" as they pass through matter, if you could vibrate the particles of an object at a rate that made it "less opaque" to the sub-quantum vortices, you should be able to reduce that object's effective mass; and thus, here on the surface of the Earth, its "weight." Is this what Leedskalnin did?

And might Tesla's "Flivver" have operated by utilizing

high-voltage AC & DC fields to divert the sub-quantum pressure around his device (similar to the light-bending "meta materials" being developed by the U.S. Military for invisibility), thereby not only reducing his vehicle's mass, but utilizing the diverted sub-quantum pressure as a means of propulsion; a form of electro-gravitational "Kung Fu?" If so; very tricky, Mr. Tesla!

It is worth noting that this is not as speculative as it might seem. Two of Burkhard Heim's students—Dröscher and Häuser—have demonstrated mathematically (and verified via experiment) that it is, indeed, possible to alter the mass of an object using electromagnetic frequencies. NASA has similarly verified that Eugene Podkletnov's anti-gravity disc does, in fact, produce a reduction in mass of objects placed above the rotating superconductor. It is a *minor* loss; but consistent and verifiable nonetheless. Perhaps exploring the sub-quantum aspects of gravity will allow such researchers to improve the effectiveness of their devices.

Chapter 5:
Magnetism

Now that we understand gravity, understanding magnetism is relatively simple, as the same fundamental process (i.e., a pressure differential in the sub-quantum vortices) is at work. This is why fluid dynamicists have found that, in the absence of unbalanced electrical charges, the equations that describe the motion of a fluid vortex also describe the magnetic lines of force that coil helically around (and at right angles to) an electrical current.

Magnetic materials act as a "force concentrator," *actively* channeling the sub-quantum vortices through their internal architecture; like holding your thumb over the end of a garden hose. Or like light through a lens. This is why magnetism is stronger than gravity: while gravity is a "passive blocking" of the ambient sub-quantum pressure, magnetic materials (and electrical conductors; but that is for the *next* chapter) actively *channel* the sub-quantum pressure; and in a very specific fashion.

While the concept of pressure differential is central to the process of aerodynamic lift, I did not deal with this concept in the previous chapter, because gravity is a *passive* process. But now that we have moved on to magnetism, which is an *active* process (similar to electricity), we have to do a bit of an "Aerodynamic Lift" primer.

Just as the pressure differential between the air flowing above and beneath an airplane's wing produces aerodynamic lift; the pressure differential between the outside of a magnetic conduction zone, and the inner "shielded" zone, is what produces magnetic "attraction."

We can see how the magnetic material (since it is akin to the "thumb over the end of the hose") may increase the *velocity* of the stream of sub-quantum vortices; but at the cost of *substance*. This is *why* pressure differential in aerodynamics produces lift: because the faster-moving area has *less substance*, albeit moving at a higher velocity. (It is like time and money: the more you have of one, the less you generally have of the other.) So whether dealing with air, water, or magnetic flux (which is simply the sub-quantum vortices being channeled *faster* by the material of the magnet); faster = "less stuff" = a system that moves in the direction of the faster-moving, lower-pressure area.[†]

Just as putting your thumb over the end of the hose does not cause any *more* water to come out of the hose (quite the opposite, in fact); it simply makes it come out *faster* (like an aerated shower head or faucet, which is designed to save water), and actually allows *less* total water to come out of the hose, contrary to what a casual observer might believe upon witnessing the increased impact of the water that is emitted. Less water (the hose with thumb over the end), less air (aerodynamic lift), or fewer sub-quantum vortices: in all cases, faster flow at the expense of *substance*.

Thus, in the case of magnetic "attraction," we actually have the sub-quantum pressure differential at work in *two* different ways: one, the typical "faster = less substance" aspect of the magnetic field lines constitutes a "substance deficit," thereby producing a zone of lower pressure; and thus, allows the external sub-quantum pressure to force the objects together. But we also have the fact that the magnetic field lines act as a *shield*, producing a zone of reduced sub-quantum pressure *within* the shielded area, which allows the external sub-quantum pressure to force the objects together. No surprise, then, that magnetic attraction (similar to electricity) is so much stronger than gravity.

So we can see that it is the *closed circuit* that is created when the magnetic lines of force from a magnet are conducted through a magnetically conductive object that causes magnetic attraction. The condensed lines of force (you can see in the images on the next page that the lines of force emanating from the end of a lone magnet are different when compared with a magnet that is conducting through a ferro/para-magnetic object) in a magnetic field that is being conducted through a magnetically conductive substance create a partially "shielded" zone that creates a sub-quantum pressure differential;

[†] Base pressure being the same (14.7 PSI for air, the garden hose valve being set at the same GPM rate regardless of whether your thumb is over the end of the hose or not, and the base level of pressure of the sub-quantum pressure), an increase in the *velocity* of the substrate *must* produce a lower *volume* of that substrate; and thus, a pressure differential in the direction of the decreased volume.

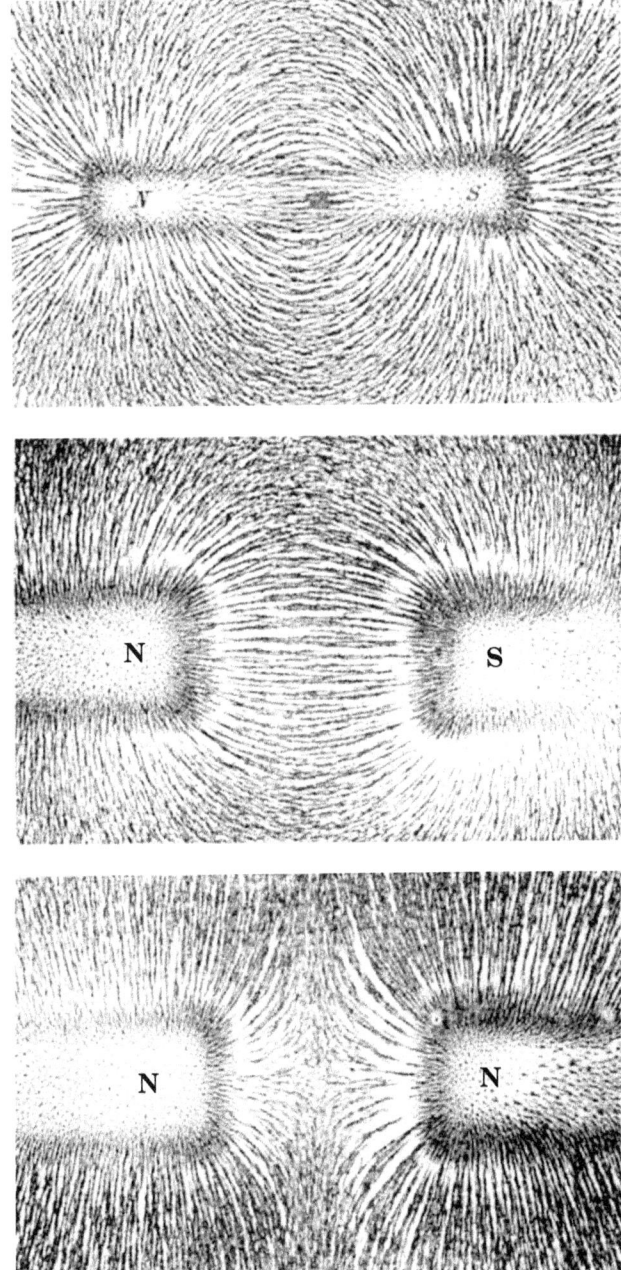

similar to the example of moving a large object in front of the fans in the previous chapter on gravity. This allows the external sub-quantum pressure to force those objects together. Thus, the denser the magnetic field lines, the higher the pressure differential between the "shielded" zone and the external sub-quantum pressure, leading to a greater force pushing the magnet and the magnetically conductive object together.

Going back to the "fans and beach ball" example in the previous chapter: imagine that, instead of a solid object, you put a large (but thin/strong) metal *frame* in the way of one of the fans, covered (as in our "boat paddle" example) with chicken wire. Were you to look down on the beach ball from above, you would likely not notice any deviation from center, as the chicken wire does not provide much resistance to the flow of air. Replace the chicken wire with the same sequence of materials used in the "boat paddle" analogy in the Gravity chapter, and you will see that the beach ball slowly progresses in the direction of the screen as the screening material becomes increasingly opaque to wind. This same effect occurs as the magnetic field lines become more dense, thus shielding the area between the magnet and the magnetically conductive object, and allowing the resulting sub-quantum pressure differential to force the objects together, just as occurs with gravity.

But while this aspect of magnetism is extremely similar to gravity (though much stronger, as it is an *active* process rather than a passive one like gravity), magnetism has an important difference: magnetism can also *repel*. And not just magnets repelling other magnets; some *materials* are actually repelled by magnets. Most people are aware that some materials are strongly attracted to a magnet (these are known as "ferromagnetic"), while other materials are only very *weakly* attracted to a magnet. (Such materials are called "paramagnetic.") But there is another class of materials (known as "diamagnetic") that are actually *repelled* by a magnetic field, much as two like magnetic poles repel each other. Magnetic repulsion is obviously a *major* difference between magnetism and gravity. How can the sub-quantum pressure differential explain magnetic repulsion?

Quite simply, actually. Look again at the diagrams depicting magnets in attraction and magnets in repulsion. You can easily see the "shielded" zone between the magnets in attraction. But do you see the odd pattern between the magnets in repulsion? What you are seeing is *excess* sub-quantum pressure that has been channeled into the area between the repelling magnets. This creates a sub-quantum pressure differential, simply in the opposite direction. Thus, with more pressure in-between the repelling magnetic poles, the magnets are forced apart by that area of greater sub-quantum pressure.

You have to remember that the very aspect of magnetic materials that makes them a "magnet" is the way in which they channel the sub-quantum pressure through the internal arrangement of their atoms. So in looking at the images of the magnetic field lines, you have to picture that what you are seeing is a representation of the sub-quantum pressure being ejected from the magnets *under great force*; just like putting your thumb over the end of a garden hose, thereby increasing the velocity of the water ejected. And what do you think would happen if you were to mount two garden hoses (say, on two wheeled furniture dollies) with the open ends facing toward each other, then turn them on; similar to the "magnets in repulsion" image? Exactly what you would expect to happen. The increased sub-quantum pressure between the repelling poles forces the magnets apart, as that pressure is greater than the external sub-quantum pressure.

To continue this metaphor, let us return to the "magnets in attraction" topic for a moment. When dissimilar poles are adjacent, it is like putting a garden hose next to a powerful "Shop Vac." Every last drop of water that is emitted by the garden hose is readily absorbed by the powerful vacuum. Not a drop hits the ground below. Imagine this creating what looks like a "rope of solid water" between the end of the garden hose and the nozzle of the Shop Vac. Now, ask yourself:

were it to start raining very lightly (i.e., not a torrential downpour that could rival the volume of water being ejected by the hose; just a light sprinkle), and the "garden hose-Shop Vac" system were only slightly suspended above the ground; how many of those rain drops do you think would make it to the ground beneath? That's right: few to none. And why? Because the "solid rope of water" between the garden hose and the Shop Vac nozzle would act as a *shield*, keeping the small raindrops from penetrating the protective layer of water. The ground beneath the garden hose-Shop Vac system should be as dry as the ground beneath a freeway overpass in a light rain. Extend this analogy to three dimensions, and you can see how the field between the magnets in attraction mode "shields" the area between them; the extent to which this occurs depending upon the density of the magnetic field lines. Instead of water shielding against other water, we have magnetic field lines (just a stream of sub-quantum vortices ejected at high velocity, after all) shielding against ambient sub-quantum vortices. This creates a zone *inside* of the shielded area that has *less* sub-quantum pressure; and all such "pressure differential" systems allow the external pressure (which has *not* been diminished) to move the system in the direction of the lower pressure. In this case, it allows the external sub-quantum pressure to force together the magnet and whatever substance through which it is conducting (be that another magnet, or a piece of ferro/para-magnetic metal).

When the sub-quantum pressure flows through materials we refer to as "magnetic" (i.e., when their internal atoms are properly aligned), we get what we call a "magnetic field."

Other materials—such as copper and silver—behave similarly (i.e., they allow the sub-quantum pressure to flow through in a different way), and we call this particular effect "electrical conduction."

When you move a permanent magnet through a coil of copper wire, that motion causes the sub-quantum pressure to flow through the copper; for as long as the motion continues. But as with a syringe full of water; stop squeezing, and the flow of water ceases.

Neither magnetism nor electricity is a separate entity in-and-of itself; merely the result of the sub-quantum pressure interacting with various materials in different ways.

We note, for example, that electric fields are always at 90-degree angles to magnetic fields. As an analogy to understand this from a sub-quantum pressure perspective, think of the sub-quantum pressure as light, and magnetism/electricity as the result of said light passing through vertical and horizontal blinds. As you change the position of the slats, you can also let through more or less light depending upon the position of the slats. In both cases, the light is still light; but in one case, it will project a vertical pattern onto everything in the room; in the other case, a horizontal pattern. These horizontal and vertical patterns are magnetism and electricity (with the intensity of the light corresponding to the strength of the magnetic or electric field): at 90-degree angles to each other; but in reality, they are nothing more than distinct patterns produced by the sub-quantum pressure being actively channeled through specific materials.

You will also note that the mainstream explanation for *why* electric and magnetic fields are at right angles to each other is lacking; they simply note that it is observably so, measure it extensively, and create equations to describe what they have measured; and then leave it at that. Just as they do with gravity. But when you consider the sub-quantum vorticular nature of the most fundamental level of reality, then factor-in "fractal expression;" one cannot help but notice that a *vortex* has an axis of rotation (current flow), and a direction of rotation (magnetic field) that is at 90-degrees to that axis.

Magnetization of Susceptible Materials

Aligning the internal arrangement of magnetic materials allows that material to channel the sub-quantum pressure in such a way as to produce a "magnetic field." But what do we *mean* by a "magnetic field." Just as putting your thumb over the end of a garden hose turns the slow, gentle stream of water that is normally emitted into a highly pressurized *jet* of water; in much the same way, highly magnetizable materials that have had their internal elements aligned properly— thereby turning them into a magnet—are like the "thumb over the hose" that allows them to emit a stream of highly-accelerated sub-quantum vortices; such that other magnetizable materials are affected by this stream. Not only do such materials *conduct* the magnetic lines of force, allowing the external sub-quantum pressure to force these materials and the magnet together; but the internal elements of the (as yet) un-magnetized (but magnetizable) material eventually become magnetized. These materials (think of the humble paper clip, for example) will not become magnetized while simply lying around (i.e., while the ambient level of sub-quantum pressure is flowing through them). But if you expose them to a strong magnetic field, the internal elements begin to align in such a way that they, themselves, begin to channel the sub-quantum pressure to a sufficient degree that *other* magnetizable materials can then be affected.

You can put electrical energy into a piece of wood via a coil of wire, and get nothing. But put that same amount of energy into a piece of iron, and you produce a magnetic field. The difference is in the internal arrangement of the material. It is all about how the ambient energy is channeled. A 100-watt light bulb lights-up a room, but cannot burn anything, unless you lay something flammable directly on *top* of the light bulb. But a 100-watt *laser?* Even a 2-watt laser can pop balloons or burn paper from across a room. So imagine what a 100-watt laser could do. The exact same amount of energy is being expended in both the 100-watt light bulb and the 100-watt laser; so what is the difference? Focus. Concentration. Coherence. Similarly, when the sub-quantum pressure penetrates wood, it produces neither electrical nor magnetic fields. But that same energy, when penetrating magnetic materials (which act as force concentrators for the sub-quantum pressure; much like a lens focusing light), produces what we call a magnetic field. It is the same amount of ambient energy that penetrates both the wood and the magnetic material; but it produces vastly different effects depending upon with which type of material it interacts.

With regard to the maximum possible strength of magnetic (or electrical) attraction (or repulsion); much like gravity, once the maximum level of screening is reached, there will be no possibility of increased pressure differential; and thus, no possibly of greater attraction, since that "attraction" is, in reality, simply the sub-quantum pressure forcing the objects together. Since the sub-quantum pressure has a fixed value (much as Earth's air pressure is 14.7 PSI)[†], there will be an upper limit to the amount of possible force pressing the objects together.

Once 100% of the sub-quantum pressure is shielded, no amount of additional energy input will increase the "attractive" force between two magnets; or between a magnet and a magnetically conductive object.

Since we have just discussed the energy in a magnetic field; what exactly *is* the source of the energy in a magnetic field? Picture two strong (rare earth) circular "donut" magnets slipped (in repulsion mode) over a wooden pole. Obviously, the top magnet will be suspended above the bottom magnet. Now, place a small weight plate (1-2 lbs.) on top of the upper magnet. While the extra weight will cause the top magnet to sink lower on the pole, you will note that it is still suspended above the bottom magnet. Now, according to the physics definition of "work," no "work" is being done when the weight plate is being held-up by the magnets.[††] By this definition, a light bulb is also not doing any "work;" but we can see that *energy*

---

† The sub-quantum pressure would not be noticeable to us, just as the 14.7 PSI of atmospheric pressure is not noticeable to us (nor is the deep ocean pressure noticeable to deep-ocean fish). The sub-quantum pressure could be truly massive, and we would never notice. Its value will determine the maximum level of gravitational pressure, magnetic "attraction," electrical potential, etc.

†† In physics, "work" is the "dot product" of force and displacement; i.e., when a force acts on a body to produce a displacement of the point of application in the direction of the force. Thus, if there is no displacement, you have multiplication by zero; which means no "work."

is definitely being expended in both cases. And if energy is being expended, what is the *source* of that energy?

We have all seen the children's toy that levitates a small foam ball using a flow of air. Since the ball is hovering in-place, suspended in mid-air *above* the "blower" apparatus, (i.e., no displacement, and therefore, no "work" being done), energy is obviously being expended. A battery-operated motor is running, expending energy in order to provide the air to keep that ball floating. But in this case, the batteries supply the energy to run the motor. In the case of the light bulb, it is plugged into a wall. But whence comes the energy for the magnet that levitates the weight plate? While mainstream physics has no sensible answer, we can see in the sub-quantum pressure explanation of magnetism a ready answer. Much as a lens merely focuses external light, a magnet is simply channeling the external sub-quantum pressure.

Occam's Razor and the Inverse Square Law

Every time we see a similar phenomenon in different places, or expressing itself in different ways, what does the scientific mainstream generally do? Why, they come up with a different explanation for each and every instance, of course; which would cause Sir William of Occam to frown.

Occam's razor states (variously) that "entities should not be multiplied unnecessarily," "It is vain to do with more what can be done with fewer," and "A plurality is not to be posited without necessity." Or as most people in modern times translate it: "The simplest solution usually is the correct one." And nature is very efficient that way; which is why we see such widespread consistencies as, for example, the inverse square law. You see it in gravity, you see it in electrostatics, and magnetics. But the standard model posits different explanations for all of these various phenomena. Would it not make more sense that, when you see such striking similarities in the behavior of (as well as the equations describing) multiple phenomena, that the same underlying phenomenon might underlay them *all?* The logic mind responds "Absolutely!"

But the Scientific Mainstream refuses to delve beneath the hallowed "Planck length," into the sub-quantum, in order to find the common thread. Only such figures as David Bohm and John Wheeler dared to tread there.

Similarly, we see this with regard to the speed of light—of course, that is the speed at which *light* travels; but then we witness such entities as electricity (whether traveling through wires, or as electromagnetic fields through the air) and magnetism *also* obeying the same speed of light limit. And yet, the Scientific Mainstream does not have an explanation; they do not even *try*. They simply conclude: "Well, light obeys it, so does electricity, and so does magnetism. Huh. Isn't that interesting. Must be a universal limit of some kind. Weird." Nobody wants to go to the next level, and look deeper.

Consider for a moment the previously discussed model of the sub-quantum pixel grid; the basic underlying sub-quantum infrastructure for the infinite universal hologram. Bearing that model in mind, understand that any magnetic coil has what is known as a "relax rate." You energize the coil; and then, when you turn the power supply off, the magnetic field inside of that coil takes a certain minimum amount of time to collapse after it has been energized; and you cannot re-energize the coil until it has been discharged. This phenomenon is well-known in the electronics industry, and is referred to as "back EMF." You have to calculate for it; put in appropriately-rated diodes, metal oxide varistors, etc., so the back EMF does not fry your circuit. But the bottom line is that it takes a certain amount of time to discharge that energy, and there is nothing you can do about it.

This is similar to the pixels on a TV or computer monitor. They have a "refresh rate" specified in Hertz. When you zap one of those pixels, the little phosphorus dot on the back of the TV screen is going to glow for "x" amount of time. Period. No matter what you do, it is going to take "x" amount of time before it goes dark again. You cannot refresh that screen any faster than the minimum discharge rate of that little phosphorus dot that is painted on the back of the screen. That is the physical limit, at least in that type of screen.[†]

The sub-quantum pixel grid's "minimum refresh rate" *is* the speed of light. That is the absolute fastest that the sub-quantum vortices can agglomerate into any of the "fundamental particles"—that can only happen so fast. Thus, everything from the "fundamental particles" on up—what we know as macroscopic "matter"—is subject to that minimum refresh rate; that minimum amount of time that is required to produce what we see as "physical movement." (i.e., The sub-quantum pixel grid ceasing to "display" something at one position, and begin displaying it at another; the illusion of motion, analogous to the "frames per second" of any video.)

To truly grasp this, you must remember that what *we* call "matter" is actually nothing but empty space.[††] And since it is nothing but empty space, it is all comprised of the sub-quantum vortices; thus, it makes sense that everything from that bottom level up is going to be subject to that minimum refresh rate; at least if you play by the rules. And that is the whole trick of it: *if* you play by the rules. If, instead, you do things like Tesla did with his "scalar waves."[†] (Scalar = sub-quantum) That technology was instantaneous, and nothing could block it: just like gravity. It was akin to a "tug on a rope"—felt instantly at the other end. In such instances, we are able to "peek

---

[†] Plasma screens, LCD screens, etc. also have refresh rates; but CRT technology has been around longer, so its principals are more well-known to the general public.

[††] Most mainstream physicists will say it is "mostly" empty space, because they still think of particles as an actual "solid particle." Now, the really *good* physicist—even in the mainstream—will admit that there is no such *thing* as a "solid particle;" they know that even the electron is a little miniature *vortex*. Actually *vortices*, though they are not yet aware of that. But we will at least give them points for even admitting that the electron is a vortex. It would be splitting hairs to point out that it is actually a combination of different vortices from most likely six different cardinal direction angles; but again, give them a cookie (at least Leonard Susskind would appreciate that—inside joke, for anyone who has attended one of his lectures) for at least getting past the "solid particle" concept. It is a big leap from "the electron is a tiny marble" to "well, it's not really *anything*." The reason most of them have a hard time with that is because when they think of a vortex, they think, "The vortex has to be in some sort of *medium*, right?" They cannot comprehend that pure angular momentum—a fundamental "tendency to spin;" *that* is the sub-quantum vortex; which is why "electron as vortex" has been taking hold *slowly*. They concede that the electron *is* a vortex; but since they do not know *why*, they are not willing to go below the Planck length. It is catching on, but it is catching on more slowly than it otherwise might if there were something they could point to and say "Oh, it's a vortex in *that*!" Like the ether—but they will not go there either.

[†††] Tesla's scalar wave communicators were first used on the remote-controlled submarine that he demonstrated for the military. He wanted them to use it for the war effort.

behind the curtain" and see that those are fundamental sub-quantum phenomena.

That is how you can distinguish between a sub-quantum phenomenon and a standard macroscopic phenomenon: the speed of light limit. This limit will apply for any macroscopic phenomenon such as electricity, magnetism, and light; but not for a sub-quantum phenomenon. So if we want to do something such as travel faster than light, you cannot do it in a physical ship. At least, not in the normal way; especially since there is no such thing as a "wormhole." But if you were to break something down at "Point A" and, instead of reassembling it at "Point A_1" immediately adjacent (i.e., in the "normal" way); you simply reassemble it at "Point Z" some distance away (since communication of *information* is instantaneous across the universe), that would be a way around the speed of light limit. That is how you have to do it: through the sub-quantum, because anytime you do it in such a way that you would be subject to the minimum refresh rate, you hit the speed of light limit. The speed of light limit is not insurmountable; it is simply insurmountable doing it the "normal" way. Just as you cannot simply "walk through a wall" in a video game (you have to obey the rules of the game), and you cannot split an atom with a hammer; it simply does not work that way: you have to "play by the rules" at the macroscopic level. If you want to "cheat," you have to work at a sub-quantum level. "Tweak the projector," so to speak.

Chapter 6:
Electricity

I very nearly rolled this chapter into the previous chapter, since so much of the material on magnetism applies equally to electricity. But rather than dealing with them both in a single chapter titled "Magnetism & Electricity," I felt that electricity still deserved its own chapter, regardless of how brief it might be. In fact, it may actually help to underscore the point: that the same fundamental phenomenon underlies gravity, magnetism, *and* electricity.

Just as magnetic materials act as a "force concentrator," *actively* channeling the sub-quantum vortices through their internal architecture; so, too, do electrical conductors (whether copper wires here on Earth, or the smallest known particles floating in space at extremely low saturation) actively channel the sub-quantum vortices through *their* internal architecture. Rather than a magnetic field, this produces an *electric* field.

And similarly to magnetic "attraction," the pressure differential between the outside of an electrical conduction zone, and the inner "shielded" zone, is what produces *electrical* "attraction." (See images below)

With electrical "attraction," we also have the sub-quantum pressure differential at work in *two* different ways: one, the typical "faster = less substance" aspect of the electric field lines constitutes a "substance deficit,"

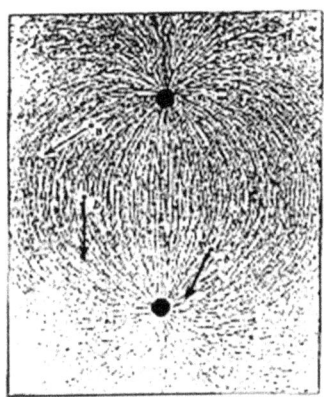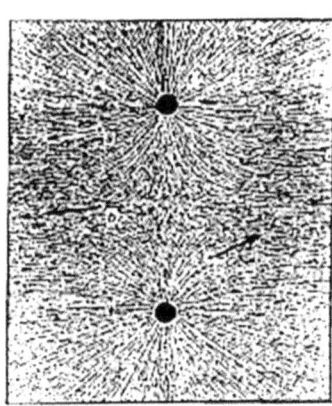

Opposite Charges Attracting   Like Charges Repelling

thereby producing a zone of lower pressure; and thus, allows the external sub-quantum pressure to force the objects together. But we also have the fact that the electric field lines act as a *shield*, producing a zone of reduced sub-quantum pressure *within* the shielded area, which allows the external sub-quantum pressure to force the objects together. No surprise, then, that electrical attraction is 10^{39} times stronger than gravity.

And once again, just as with magnetism, it is the *closed circuit* that is created when oppositely-charged electric field lines interact that causes electrical attraction. The condensed lines of force (see images on previous page) between two oppositely charged objects create a partially "shielded" zone that creates a sub-quantum pressure differential. This allows the external sub-quantum pressure to force those objects together. Thus, the denser the electric field lines, the higher the pressure differential between the "shielded" zone and the external sub-quantum pressure, leading to a greater force pushing the electrically charged objects together.

In the case of electrical *repulsion,* if you look at the diagram of *similarly* charged objects, what you are seeing is *excess* sub-quantum pressure that has been channeled into the area between the repelling electric fields. This creates a sub-quantum pressure differential, simply in the opposite direction. Thus, with more pressure in-between the similarly charged objects, the objects are forced apart by that area of greater sub-quantum pressure.

You have to remember that the very aspect of electrically conductive materials that makes them a "conductor" is the way in which they channel the sub-quantum pressure through the internal arrangement of their atoms. So in looking at the images of the electric field lines, you have to picture that what you are seeing is a representation of the sub-quantum pressure being ejected from the conductors *under great force*. The increased sub-quantum pressure between the repelling poles forces the charged objects apart, as that pressure is greater than the external sub-quantum pressure.

Realizing now the identical underlying phenomenon that underlies gravity, magnetism, and electricity, we can see that neither gravity, magnetism, nor electricity is a separate entity in-and of itself; merely the result of the sub-quantum pressure interacting with various materials in different ways; both passively (in the case of gravity) as well as *actively* (in the cases of both magnetism and electricity).

Now, we have to understand that, since the ambient sub-quantum pressure is ubiquitous and isotropic, it is a "zero sum" game; until the clusters of sub-quantum vortices we call "matter" get in the way, and *interact* with the sub-quantum pressure. This gives us gravity, magnetism, and electricity. The point here: it is only when the *equilibrium* is disturbed that we see "interesting" or "useful" phenomena. This leads to the bi-polar effects we see in magnetism and electricity. Since gravity is a more *simple* result of a *passive* pressure differential, we see only one "pole," as this is the most basic result of a pressure differential. Magnetism and electricity both result from more complex, *active* interactions between the sub-quantum pressure and the internal arrangements of various materials; like the difference between water flowing through a smooth, perfect, poured concrete channel; and water flowing through a natural stream, with rocks along the bottom producing whirlpools, larger rocks producing "white water rapids," logs in the way, bends in the stream, silt on the bottom, etc.

Let us consider the specific case of the basic phenomenon utilized to create an electrical generator. As a magnet moves through a coil of wire, picture its magnetic field acting as a "cow catcher" on a train, pushing the sub-quantum vortices along through the wire. Whereas the copper *was* at equilibrium—"an object at rest;" equal sub-quantum pressure from all sides—when you "comb" through the copper with a magnetic field, you create a disequilibrium; a vacuum that the ambient sub-quantum pressure immediately rushes-in to fill. We call this "current flow." As soon as you stop the physical motion, the current also stops.

This is why ferromagnetic cores inside of copper coils increase the effect: the magnetic lines of force become more concentrated, thereby increasing the disequilibrium; and therefore, producing more current.

Of course, there are other ways to produce this disequilibrium (chemically, for instance; as in a car battery); but no matter how this disequilibrium is produced, it is required for current to flow.

It should be apparent from this example that what *we* think of as "electricity"—current flow in circuits, motors, etc.—is simply the way that we have *engineered* this basic underlying phenomenon. In reality, any time you have a magnetic field, there is an electric field at right angles to it. And what is an electric field? An accelerated stream of sub-quantum vortices, exactly as with magnetism. But what we utilize for the purposes of our devices is the *charge separation* aspect. A magnetic field has a North and a South pole, while an electric field has a Positive and a Negative pole; and we utilize this charge separation to produce devices that perform work for us.

At its most fundamental level, an electric field is the same as a magnetic field; produced in the same way (sub-quantum pressure channeling), and with the same separation of equal-but-opposite poles. The fact that we manipulate this effect in various ways to drive our technology at the macroscopic level is neither here nor there; the fundamentals are the same.

We can see now that the three most prominent phenomena of reality—Gravity, Magnetism, and Electricity—are all produced by the same underlying phenomenon: the ubiquitous and isotropic sub-quantum pressure.

Considering electricity specifically, if we could find a way to induce the disequilibrium effect at-will, we could have all of the free energy we needed. Similarly with gravity: if we ever develop the ability to directly manipulate the sub-quantum pressure, this would allow us complete control over gravity.

Another area that would be positively impacted by this understanding of the basic phenomenon underlying electrical current flow is superconductivity. To understand the difference between normal electrical conductivity and *super*conductivity, go back to our example of water flowing: think of water trying to flow in a river that is clogged-up with branches, rocks, etc.—a chaotic mess; meaning that the flow of water will be restricted. Now, picture a river that is completely *free* of debris: how much more smoothly the water will flow. In a normal electrical conductor, random molecular motion in the atoms of the material restricts the flow of electricity through the conductor; similar to how trying to walk down a busy city street—getting bumped-into by various people—restricts your ability to make good time toward your destination. But in a *super*conductor, the random molecular motion has been suppressed by cooling the conductor to extremely low temperatures. This is like the water flow in a "clean" river; or walking down that same city street with *no* people obstructing your forward motion.

But, while current techniques of superconductivity do *work*, they are not very practical (which is why the hunt for a "room temperature superconductor" is the "holy grail" of research in this field), as mainstream science is coming at the problem from the wrong angle. Rather than deal with the underlying *source* of the problem, they are dealing with the *symptom*. And until they cease ignoring the sub-quantum, they will be relegated to treating symptoms only. This is why Nikola Tesla was so successful in his endeavors: he did *not* ignore the sub-quantum. On the contrary, he *based* his research upon the sub-quantum as a *given*. His success in his work speaks volumes for the efficacy of this approach. And until the scientific mainstream is willing to do likewise, and develop a technological infrastructure capable of functioning on a sub-quantum level, they will be barred from achieving similar success, keeping such accomplishments as room temperature superconductors out of their reach. Any progress they make in this area will be slow and painful at best, as they will be relying upon "luck" and trial-and-error for every small advance. Quantum leaps in the generation of electrical power and superconductivity could be within their grasp if only they would consider the sub-quantum underpinnings of gravity, electricity, and magnetism. Then, and *only* then, will they succeed in (as Tesla put it) "attaching their machinery to the very wheelwork of nature."

Chapter 7:
Strong & Weak Nuclear Forces

No discussion of physics would be complete without mentioning the final two purported forces: the so-called "strong nuclear force" and "weak nuclear force."

Right out of the gate, we can dismiss the "weak nuclear force," since it has been shown (via the discovery of the "electroweak interaction," for which a Nobel Prize was duly awarded) to be nothing more than a manifestation of electromagnetism; leading some theorists to question whether they might both be manifestations of a more fundamental force. Currently, researchers are attempting to prove that gravity and electromagnetism are also simply different manifestations of the same fundamental force. Sounds a lot like the sub-quantum vortices to me. Tesla, Bohm, and Wheeler would be proud.

While physicists become perplexed upon witnessing a particle emit another particle that is more massive than the emitting particle (as occurs in "beta decay," which the "weak nuclear force" was originally concocted to explain); we can readily see—since we know that mass is an illusion—that the sub-quantum pixel grid sees a bar of gold as being the same as "empty space." Just as a TV does not care what its pixels display. One holographic projection is the same as any another. Since "mass" is an illusion—everything is really just "empty space"—we can see that the only aspect of any real import is the *information*. So, when a particle with very little mass sends out the *information* for a much more massive particle to be emitted immediately adjacent to it; nothing could be simpler for the sub-quantum pixel grid. Mass is an illusion perpetrated by specific arrangements of internal architecture in "holographically projected" particles.[†] It is similar to putting your hand out the window of a moving vehicle: position your hand horizontally, and it does not feel like much force is pushing against your hand. But position it *vertically,* and you feel quite a lot of force pushing against it. Did the amount of wind change?

---

† It is difficult to miss the implications for transmutation. Once we have created a technological infrastructure capable of directly manipulating the sub-quantum vortices, not only will we be able to easily transmute any element into any *other* element; we should also be able to "create" matter out of (seemingly) "thin air." If you are a "Star Trek" fan, think "replicator."

No. Did the mass, size, or composition of your hand
change? No. But simply by changing your hand's *position*
relative to external factors, you perceived the force of the
wind in quite a different way. This is what happens when
any macroscopic matter (matter itself being comprised
of nothing but agglomerations of sub-quantum vortices,
after all) interacts with the sub-quantum pressure: the
internal *arrangement* of the macroscopic matter dictates
precisely *how* it interacts with the sub-quantum pressure,
as well as how it interacts with other macroscopic matter.
So you can have a particle that seems to have "low mass"
suddenly emit a particle that has "high mass"—simply the
hand turning face-on to the wind rather than sideways.

Similarly easy to explain is the so-called "strong
nuclear force." Anyone who has studied the "Standard
Model" of particle physics quickly notes a penchant for the
scientific mainstream to invent a new "force" or "particle"
at the drop of a dime. Often, this is simply because they
refuse to look below the Planck Length, and recognize
that there is a sub-quantum level underpinning reality.
But in the case of the so-called "Strong Nuclear Force,"
well-known physicist Richard Feynman's "like-likes-like"
explanation does the job nicely.[†]

You will note, upon close examination of the Periodic
Table, that there is never an element with more than one
proton in its nucleus without an equal or greater number of
neutrons. (With a "neutron" being, in reality, nothing more
than a proton that has merged with an electron. Picture
the electron's wave pattern—as we discussed in the
"Waves vs. Particles" chapter—wrapped around a proton,
and you will get the idea.) This is why the neutron is

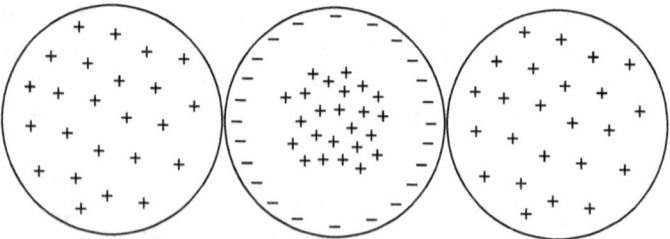

† An illustration of Feynman's "like-likes-like" effect

very nearly electrically neutral: its positive charge is being neutralized by the electron's negative charge. Remembering the inverse square law that applies to electricity (as well as gravity and magnetism), we can see that, with the neutron's exterior "coating" of the negatively charged electron wave pattern, positively charged protons would be attracted to the neutron. The proximity of the positively charged proton drives the concentration of positive charge in the neutron deep into the center of the neutron, thereby increasing the negative charge on the surface of the neutron, and binding it even more powerfully to the proton. This is why the scientific mainstream mistakenly believes that the "strong nuclear force" is more powerful than electromagnetism: they are calculating how much force would be required to force two like-charged protons together. But when you combine the effects of the inverse square law with the charge migration that occurs within the center of the neutron when it is in proximity to a proton; we find that no such "strong nuclear force" is required. It is, as Feynman so aptly dubbed it, simply a case of "like-likes-like." Feynman's hypothesis has been verified by experiments conducted by Norio Ise at Kyoto University. (Nagornyak et al., 2009)

Chapter 8:
Sub-Quantum Holography

Before we transition to "Part Two," we need to cover the basics of holography. Many of you are likely familiar with this subject (especially those who have already read Talbot's "The Holographic Universe"); but for those who are not, this is an essential step to understanding the material in "Part Two." And even for those who may already be holographic experts, the information in this chapter regarding the tie between holography and the sub-quantum will be crucial.

The basic essence of a hologram is actually quite simple: it is an interference pattern. When two otherwise identical sources of monochromatic light are allowed to expose a piece of film, with one of the sources having first been allowed to reflect off of an object, the interference patterns that are captured on the film are a record of data describing that object. The basic configuration of a laser interferometer for holography is depicted below:

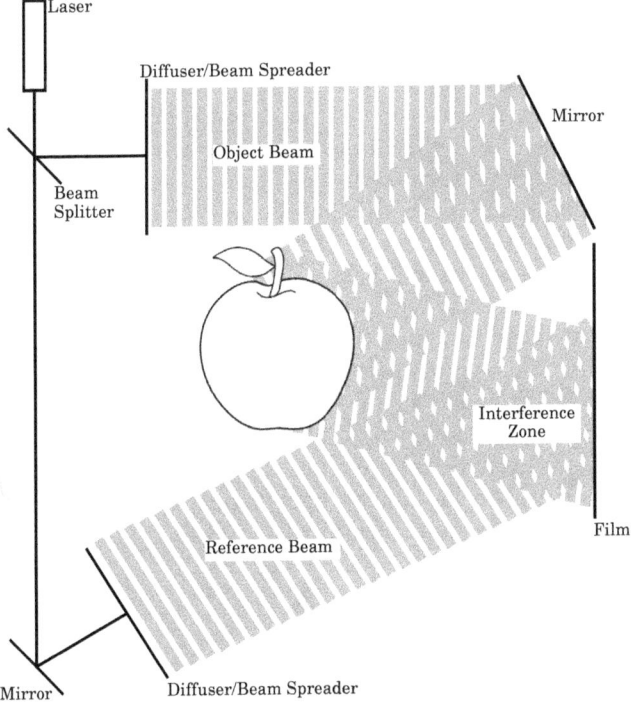

Once the film is developed, the holographic data contained therein can be expressed by passing a laser of the same frequency as the original beam (ideally passed through a similar diffuser/beam spreader) through the developed film. This will cause a three-dimensional representation of the original object to emerge from what appears to the human eye to be a piece of film full of nothing but "squiggles"—interference patterns.

But the interesting part is that, while the human *eye* may perceive the holographic film to be nothing but "squiggles," the human *mind* has no trouble deciphering these interference patterns as the original object. In 1979, Drs. Russel & Karen DeValois of UC Berkeley discovered that cells in the human visual cortex contain "feature detectors" that respond to Fourier translations of images, not to the original images.[†]

As an example: you can show someone a Fourier image of an apple; and while that person will consciously only see interference patterns (such as you see in a holographic film), their brain responds as if they had seen an actual apple. Just as your retina sees the world upside down, but your brain flips it right-side-up for the sake of convenience; your brain recognizes the interference patterns that constitute the true holographic infrastructure of reality, while constructing a "GUI" ("Graphics User Interface," for the *non*-"computer geeks") for your convenience. (More on that in "Chapter 10: Holographic Sub-Quantum Mind.")

Utilizing the fact that multiple holograms can be recorded on a single piece of film by changing the angles between the lasers and the film for each image, one square inch of holographic film could hold over 60,000 pages of text.[††] So imagine how much data can be stored in the infinite, universal holographic storage medium that is the sub-quantum "pixel grid."

[†] For the *non*-"math geeks" out there; Fourier translations are the mathematics of holography: the mathematics that convert visual images into wave forms.

[††] Considering that a 650-Meg CD-ROM can hold roughly 100,000 pages of text; fitting 60% of that onto a one square inch piece of film is quite impressive; especially when that data can all be read near instantaneously.

As an analog to the different angles used to store multiple separate holograms on one piece of film in laser-based *film* holograms; think of different *phases* (of which there are a near infinite variety) being used to store multiple "angles" of data in the infinite, universal sub-quantum holographic storage medium.

This illustrates that, where *we* may perceive something as relatively "small" (like one square inch of holographic film), a tremendous amount of data can be stored in that "tiny" space. Just as computer storage form-factor CDs went from 650-Megs to now more than 100-Gigs for the latest Blu-Ray format; all on the same size as an original CD. Modern Blu-Ray 100-Gig readers/writers can also read/write 25-Gig and 50-Gig Blu-Ray discs, as well as DVDs and CDs. All of these disc formats utilize the same form-factor, and *similar* technology; but they have drastically different data storage capacities. And holographic data storage technology ups the ante even *further.*[†] As we continue to develop our technological infrastructure and encoding methodologies, we will see ever-higher data storage density. Until our technology finally reaches the sub-quantum level, and we realize that there is truly *infinite* storage capability in the universal sub-quantum holographic storage medium; with instantaneous information transfer across infinite distances. Quite a lot to get one's mind around.

One of the problems that well-respected physicists such as Bohm and Wheeler encountered when promoting the idea of a sub-quantum level of reality, was the association in the minds of the scientific mainstream of "sub-quantum" with "ether." Back in Tesla's day (toward the end of Tesla's life, in fact), when the concept of an "aether" (or "ether") began to fall out of favor in mainstream science, the primary objection was that such a medium would have to be near

---

[†] The Optware company is currently working on a consumer model HVD (Holographic Versatile Disc) drive that will store 1-Terabyte of data on a disc the same size as a Blu-Ray disc. And in stark contrast to the slow read rates of standard CD/DVD/Blu-Ray media, HVD data access is incredibly fast. See http://electronics.howstuffworks.com/hvd.htm for further information.

infinitely rigid if it were to be capable of conducting particles at the speed of light; and yet, such a medium would have to allow matter to pass through itself with no resistance. Unable to resolve this apparent contradiction in their minds, the scientific mainstream simply dropped the concept of an "ether" altogether. Had they instead followed Aristotle's advice—to examine their premises, and discover the faulty assumption they were making—science today would be decades ahead of where it now stands.

When we realize that, rather than the mythical "ether" being a *medium* through which gross matter moves (this was the faulty assumption that was never adequately examined when the scientific mainstream moved away from the concept of an "ether"), and via which electromagnetic waves are conducted; in reality, what Tesla and his contemporaries would have called the "ether" is actually part of the "projection system" for our holographic universe. (See chapter two)

With the sub-quantum pixel grid, there is no need for a paradoxical "ether" possessed of impossible to resolve properties, as matter is simply a "projection" of the pixel grid. And just as a TV's pixel grid can display something as fast as light, or as slow as a snail; so, too, the sub-quantum pixel grid can "project" anything, anywhere, at any rate of speed (up to the limit if its refresh rate, as we discussed previously); and yet, it does not "interact with matter," as matter is simply another "projection." And, being based completely upon *interference patterns,* we can now see that it is a *holographic* projection system. "As above, so below." Fractal expression, once again.

The double-slit chapter showed us that, at the sub-quantum level, the universe is all about interference patterns. Combine this with the "fractal expression" effect, and we can see that the universe cannot *help* but be holographic. (As Michael Talbot showed so capably that it *is* in his book, "The Holographic Universe.") Going further—realizing that the infinite, universal sub-quantum "pixel grid" constitutes a truly infinite holographic information-storage medium—this actually allows us to show *why* the universe is holographic. Since every smallest unit in the incompressible sub-quantum pixel grid is in direct contact with every *other* "pixel" in the universe (like a table full of intermeshed gears: turn one, and they *all* must turn), the transmission of information is therefore instantaneous across the entire infinite universe. Thus, we can see that every action of every entity in the universe is instantly stored in this infinite holographic storage medium (so well illustrated by David Bohm with his "ink drop in glycerine" example; what he called "enfolding" the data), in much the same way as the interference patterns used to create film holograms record the three-dimensional data for an object onto a piece of film.[†]

This is *why* the mind reacts to Fourier translations of objects as if they were the object itself: because all any so-called "object" *really* is, is an interference pattern: a hologram. That, of course, includes *us*.

[†] At least, for its *exterior* appearance. If we were to utilize a more penetrating form of electromagnetic wave—such as microwaves—we would be able to record (providing a recording medium that could record microwaves) three-dimensional *internal* data for the object as well. But that is a subject for a later chapter.

Part Two:
Holographic
Sub-Quantum Mind

Chapter 9:
Mind Outside Of Body

"My brain is only a receiver. In the universe there is a core from which we obtain knowledge, strength, inspiration. I have not penetrated into the secrets of this core, but I know that it exists."
                                        — Nikola Tesla

Having laid the foundations of sub-quantum *physics* in the first half of the book, we now come to the meat of the matter: the actual sub-quantum, holographic nature of the *mind*. But before we plunge headlong into that subject in the next chapter, we must first demonstrate that the mind exists *outside* of the body, and not simply within the wrinkly folds of our gray matter.

In his book "Supernormal," Dean Radin discusses an argument that occurred between neuroscientists and the Dalai Lama regarding the true nature of consciousness. When the Dalai Lama stated that the mind is "separate from matter," and not an epiphenomenon of neurological processes within the physical brain, neuroscientists railed against the "spooky" nature of such a claim. But as we have shown in the first half of this book, there is a scientifically understandable infrastructure that allows us to explain how this is possible, with no "spookiness" required. Hopefully, *Holographic Sub-Quantum Mind* can end this debate once and for all, finally uniting both camps in this age-old argument. After all, the "feelings of oneness" that have always been a large part of the teachings of most meditative disciplines are to be expected in a *holographic* universe, where the minds of *everyone* are *all* distributed across the same infinite holographic storage medium.

In the May, 2012 edition of "Physics Essays," Dean Radin published the results of an experiment wherein a group of volunteers were asked to visualize *blocking* the left slit in a double-slit experiment; and the interference pattern actually showed a reduction. The adept meditators were able to produce a more pronounced disruption of the interference pattern than non-meditators. Since we have already shown (chapter one) that the double-slit experiment is a result of interference on a sub-quantum level, the fact that people's minds can influence the double-slit interference pattern is evidence that the human mind exists on a sub-quantum level.

Electromyograph (EMG) studies of human muscle reactions show that the muscles react to external stimuli *before* the brain becomes aware of them. People who can see auras (verified by Kirlian photography, as well as EMG readings) report the same phenomenon: the body's aura reacts *first*, followed by the muscles, and *then* the brain. This clearly places the location of the mind *outside* of the body. Not surprising, since the universe is holographic; meaning that our bodies *and minds* are distributed across the infinite sub-quantum holographic infrastructure of the universe. While we remain mired in what we *perceive* as "physical matter," we may *seem* to be localized: but that is merely an illusion.

Dr. Valerie V. Hunt (a neurophysiologist) verified that the body's external electrical field[†] reacts to stimuli before the body and the brain. When electromyographic readings are taken simultaneously with EEG readings, the EMG readings show that the human body's external energy field reacts to stimuli *before* the EEG registers brain activity. This verifies the above referenced statements of aura readers.

When Dr. Hunt was asked what significance she believed these results hold, she stated: "I think we have way overrated the brain as the active ingredient in the relationship of a human to the world. It's just a real, good computer. The mind's not in the brain. It's in that darn field."

Perhaps not contained *within* the body's external electromagnetic field; but that field is definitely the body's *link* to the holographic sub-quantum mind.

At the University of Edinburgh, Scotland, in 2004, a telepathy experiment showed that the *receiver* was reacting to light stimulation before the sender. The researchers' conclusion—precognition. But might not it rather be due to the fact that *both* minds are distributed across the infinite sub-quantum holographic infrastructure? A telepathy experiment is all about the *focus* of the mind, with the receiver actively seeking

---

[†] Dr. Hunt was the first doctor to use a telemetric electromyography instrument, designed and built by NASA. It was similar to the telemetric units NASA used for astronauts, but designed to detect and relay EMG data rather than heart rate, blood pressure, galvanic skin response, etc., as NASA's units did. Working with eight of the world's top aura readers, Dr. Hunt utilized the device to verify the accuracy of their readings, finding that they were extremely accurate with the primary colors, slightly less so with the secondary colors, and not at all accurate with "subjective" colors (e.g., puce, chartreuse, turquoise, etc.) that were a blend of other colors.

to mentally connect with the sender; i.e., the sender's location. In this case, an experiment involving stimulation produced by a source of light.

We know that the *information* for the light stimulation is disseminated *instantly* via the sub-quantum pixel grid; but the light *itself* is limited to propagating at the "speed of light." Thus, the holographic sub-quantum mind of the receiver obtains the information *before* our macroscopic instruments detect the light itself. Just as Dr. Hunt's EMG studies showed that the human energy field reacts to external stimuli *first,* followed by the muscles, and *then* the brain; we would fully *expect* the receiver in the telepathy experiment to receive the *information* about the light that was about to be emitted *before it was actually emitted!*† Thus, it is no surprise that the receiver reacts to the light before the EEG of the sender has registered it. It is not precognitive; simply a response on a different level: the sub-quantum level within which the mind is actually contained.

Further proof of the existence of mind *outside* of the body comes from a medical procedure known as "Deep hypothermic circulatory arrest," wherein the patient is put into a state of hibernation at 53°-64° F. Breathing, heartbeat, and brain activity cease for up to one hour. All blood is drained from their body in order to eliminate blood pressure, and they are considered clinically dead for the length of the operation. And yet, many patients have reported accurate perceptions of events taking place in the room during such operations. Since there is no brain activity—no blood in the brain, even—during such an operation; if memories *were* stored in the brain via some sort of electro-chemical process, there would be no way for this to occur, since chemical and neural activity have been suspended. Clear evidence that the physical brain is not the seat of the mind.

---

†His mind *is,* after all, embedded within the same sub-quantum holographic infrastructure as the equipment that will be producing the flash of light; and thus, his mind will be privy to the cascade of events leading-up to the flash actually being emitted. Anyone familiar with electronics and/or photography knows that a whole host of events—including the charging of a capacitor—leads-up to the actual discharge of a bright flash of light.

The radio in your car receives broadcasts. If it quits working, you get it fixed or replaced, and then you can once again receive broadcasts. In this day and age, nobody believes that the radio itself is responsible for the broadcasts it receives: we all realize that the radio is merely a receiver; that the content it plays is received from elsewhere. Unfortunately, modern theories of human consciousness continue to perpetuate the notion that our brains are somehow responsible for the contents of our minds.

In a holographic universe, there is no such thing as "here" or "there." The idea of "location" is an illusion. Much as the data in a light hologram is distributed across the entirety of the film; so, too, is everything in our holographic universe distributed across the entirety of the infinite sub-quantum storage medium; including our minds.[†] Or as Michael Talbot so succinctly put it in *The Holographic Universe*—"Everything is ultimately nonlocal, including consciousness."

[†] Every piece of a hologram contains the whole; so, in reality, our mind *and* body are distributed across the infinite sub-quantum holographic storage medium. But speaking from our localized perspective (the elements of the "video game" interacting with each other, pretending we are solid; "suspending disbelief" so we can enjoy the "game/movie" we call "reality"), we consider our bodies to be "real" because we can "feel" them. But even within this "suspended disbelief" context, our *minds* are still distributed throughout the infinite sub-quantum storage medium; i.e., *not* contained within our fictional "meat suits."

Chapter 10:
Holographic
Sub-Quantum Mind

In the beginning of the previous chapter, I promised that, concerning the actual topic of the holographic sub-quantum mind, we would "plunge headlong into that subject in the next chapter." This chapter constitutes the last of the "foundation" chapters, after which we can begin to explore some of the applications. So without further ado, let us by all means commence.

The research of Drs. Russel & Karen DeValois of UC Berkeley shows that the human mind is a complex frequency analyzer, and can recognize recorded interference patterns (such as the films that are used to produce holographic projections) as the original object that made them.[†] Since our *minds* exist as part-and-parcel of the sub-quantum infrastructure—stored in the infinite, universal holographic "storage medium," almost as if they were a computer program, running on an infinite universal computer system (and as such, blessed with infinite processing power, as well as unlimited RAM and storage space)—what we call "physical reality" is simply a *simulation*. So when we discuss how the sub-quantum pixel grid "projects" reality, understand that this does not mean it does so in a literal "physical" sense; it simply contains the encoded data that our holographic sub-quantum mind utilizes to experience the simulation.

When we combine this with the information in the previous chapter, which demonstrated that cognition takes place outside of the body (placing the physical body and brain squarely in the role of "radio transceiver"), we can conceive of the mind-body system as a "thin client" computing system. The physical body and brain serve as the "dumb terminal," connecting to one's individual holographic

---

[†] In 1979, Drs. Russel & Karen DeValois of UC Berkeley discovered that cells in the human visual cortex contain "feature detectors" that respond to Fourier translations of images, not to the original images. (For the *non*-"Math Geeks" out there; Fourier translations are the mathematics of holography: the mathematics that convert visual images into wave forms.)

As an example: you can show someone a Fourier image of an apple; and while that person will consciously only see interference patterns (such as you see in a holographic film), their brain responds as if they had seen an actual apple. Just as your retina sees the world upside down, but your brain flips it right-side-up for our convenience; your brain recognizes the interference patterns that constitute the true infrastructure of reality, while constructing a "GUI" ("Graphics User Interface," for the non-"computer geeks") for our convenience and ease-of-use.

sub-quantum mind (the "hard drive" in this scenario), with the physical body's unique resonant signature (a combination of one's DNA and neural pattern in the physical brain) acting simultaneously as an "encryption key" and "tuning circuit," to ensure the maximum possible secure connection between body and mind.

The body's bio-energy field (the "aura", Orgone, Chi, Kundalini, Prana—whatever one chooses to call it) acts as the "antenna," with the holographic sub-quantum mind (the "hard drive") storing, not only your memories (data), but also the "programs" that comprise your personality.

While the analogy of a "hard drive" connected to the physical body in a fashion similar to a radio transceiver is convenient, understand that this is not to be taken literally; i.e., the holographic sub-quantum mind is not an actual physical construct, but simply a "whirlpool in the river."[†] It may have its own unique, individual characteristics, but it is part of the infinite sub-quantum infrastructure, in much the same way that a whirlpool is simply a unique arrangement within the water of a river, and not a separate "thing." You are able to access your own, individual holographic sub-quantum mind strictly by virtue of your unique resonant signature,[††] which synchronizes with your holographic sub-quantum mind in the same way that a guitar string will resonate when its corresponding note is played on a piano; or the way a professional singer can cause their voice to resonate with (and shatter) a glass.

[†] Individual consciousness is similar to a whirlpool within the water of a river. Our holographic sub-quantum minds are discrete constructs within the fabric of the universe's infinite sub-quantum infrastructure, in much the same way that a whirlpool is a discrete construct within the water of a river. The whirlpool is contained within the river, and is composed of the same substance as the river, but maintains a discrete identity due solely to its arrangement within the river. A block of granite or marble can be sculpted by a master sculptor, and become a work of art to our subjective senses; but it is, nonetheless, still a block of granite, albeit with slightly less mass than before. The metamorphosis is, at the most basic level, simply an illusion. This does not in any way diminish the uniqueness or separateness of the individual consciousness, any more than a sculpture is diminished by realizing that it is, fundamentally, a block of granite: it simply allows us to understand its relation to the rest of the universe. At its most fundamental level, everything is all just "one big river."

[††] For the more technically-minded, think of this as a PCAR (phase-conjugate-adaptive-resonance) system, combined with the type of multiplexing utilized with fiber-optic networks.

Dr. Karl Pribram (one of the first researchers to realize that the brain operates holographically) explained that the interference patterns created when the waveforms resulting from neuronal discharges interfere with one another constitute, essentially, a "thought hologram." Once again, we see fractal expression, as higher-order processes emulate the fundamental sub-quantum infrastructure of the universe. Our minds truly do operate in a holographic fashion, even here in our day-to-day "physical" world.

Extending Pribram's work, we can see how the holographic interference patterns within the physical brain allow it to resonate with our holographic sub-quantum mind.[†] The physical brain's unique holographic signature (in combination with our DNA, since the body itself is also a holographic construct) acts as the "key" that (ideally[††]) only fits one "lock"—that of our unique, individual holographic sub-quantum mind. This limits the extent to which overlap occurs between the thoughts, memories, and personalities of individuals. An individual's holographic sub-quantum mind may be distributed throughout the infinite sub-quantum infrastructure of the universe; but the specific "holographic sub-quantum signature" of that individual's mind/body system is the only "key" that can access it, much as a laser of a specific frequency is required to express a hologram that was created by an identical laser.

Modern research has been able to associate memory storage to specific chemical signatures; but the capacity of the chemically stored data is incapable of accommodating the amount of data contained in a memory. This admittedly confuses even the top memory "experts," but they continue to cling to the belief that these memory-

---

[†] The brain is simply a 2-way *"radio"*—if you "tweak the knobs," of *course* you are going to receive different "broadcasts." But that does not mean the broadcasts are built-into the radio. The very fact that people have had *large sections* of their brains removed with *zero* memory loss shows that the physical brain does not contain memories: they are stored "elsewhere," with the brain acting as a transceiver. Dementia/Alzheimer's is when the "radio set" begins to malfunction, and no longer pulls in signals as well as it used to.

[††] See "Chapter 12: Reincarnation, Mediumship & Possession" for the exception to this rule.

related chemical signatures are somehow "proof" that the memories themselves are being stored chemically *in toto* within the brain.

In fact, what is being stored are merely *index pointers* to the actual memories stored within the infinite holographic sub-quantum mind. Think of "shortcuts" or "aliases" on a computer desktop: they merely *point to* the actual file. Or consider a computer back-up CD/DVD cataloging program: how on earth would you *ever* find anything on piles and piles of computer backup discs without some kind of indexing/cataloging system? The chemical signatures of memory are exactly that: pointers that indicate *where* within the infinite holographic sub-quantum mind storage bank the associated memory is stored.

This is why, with Alzheimer's, a person quite often cannot remember what they had for breakfast, or what their youngest grandchild's name is; but they can remember details of thirty- or forty-year-old events. It is not their *memory* that has "gone bad," only their ability to create stable index pointers, making information much more difficult to locate.

Which brings us to the prickly topic of memory erasure; whether due to accident, age-related deterioration, or intentional erasure via EDOM (Electromagnetic Dissolution Of Memory) and similar governmental/military techniques.†

Since all that is stored in the physical brain is a *pointer* to the specific memory, the *pointer* can be erased, but the actual memory to which it pointed is still safe and sound within the holographic sub-quantum mind. Think of deleting a shortcut/alias from your computer's desktop: it does not delete the file to which that shortcut/alias was pointing: merely the pointer itself. You can still search through your hard drive, find the original file, and create a *new* shortcut/alias.

---

† "Memory erasing" techniques often fail over time (i.e., memories start to come back eventually) because the data comprising the content of a memory is not actually stored within the physical brain. Thus, all the so-called "memory erasing" techniques can do is impede a person's ability (to varying degrees, and always for a limited time) to *access* specific memories.

Could the same be done for memories? Absolutely. The "hard drive" in this case—the holographic sub-quantum mind—is vast; and thus, locating specific pieces of data (memories) could be difficult. But once a memory is located, a *new* pointer can be created, effectively restoring the memory to that individual.[†]

In addition to the index pointers, *some* memories truly *are* stored chemically within your physical brain. The difference between your *main* repository of memories (the holographic sub-quantum mind), and those that are stored chemically within your physical gray matter, is not only the fact that the locally-stored chemical signatures can be erased as easily as the index pointers, but that the memories that are physically stored in the brain are only *copies* of those that are accessed most frequently; like the *cache* in a web browser. This speeds retrieval of memories that are frequently accessed, in much the same way that a web browser's cache speeds the loading of frequently-accessed web pages. The mechanism responsible for this "cache effect" in the physical brain is similar to the phenomenon of "burn-in" on a computer monitor or plasma TV; where something that is displayed repeatedly "burns-in" to the screen. (Not a good thing in the case of computer monitors or plasma TVs, which is why "screen savers" were originally created.) Similarly, thoughts and memories that are accessed frequently from the holographic sub-quantum mind leave a residual "burn-in"[††] within the physical material of the brain. This is why there is a difference (as Jose Silva taught in *The Silva Mind Control Method*) between your "memory" of an event, and your "Alpha Recall" of that same event: the "Alpha Recall" is direct access to the data stored in your holographic sub-quantum mind, while what we

[†] It is like re-building the database on a hard drive when it becomes corrupt: the data is still on the hard drive, but the system is having trouble accessing it due to a corrupted database file. Once the database is re-built, the data can then be accessed again. One day, when the scientific mainstream develops a technological infrastructure, not only for sub-quantum physics, but for directly accessing the holographic sub-quantum *mind*, we will be able to "re-build the database" of the human mind, just as we do for an ailing hard drive.

[††] Actually a holographic resonant signature, which is why researchers such as Dr. Karl Pribram have noticed that memories are stored holographically, even within the *physical* brain. There's that *fractal expression* thing again.

generally term "memory" is usually the "burn-in" from that same event[†]—the locally-cached "copy" in the case of frequently-accessed memories, and locally-stored "index pointers" for those less frequently accessed; i.e., memories that left enough of a minor resonant signature to allow us to locate the original within the holographic sub-quantum mind, but not enough to be locally stored.

Hypnosis can assist a person in directly accessing the holographic sub-quantum mind. This is why hypnotic regression can often allow a person to access data that was not consciously "remembered"—it was never "burned-in," because the person took no *conscious note* of those details. But the details were nevertheless stored in the holographic sub-quantum mind, as they *were observed* by our senses. This can allow investigators to retrieve license plate numbers and other important details that witnesses were unable to consciously recall.

The well-known phenomenon wherein a smell (or other sensory stimulus; but smell is the most common) triggers an associated memory can easily be understood from a holographic perspective. There is a technique wherein *each* of the halves of the split laser beam that is used to produce a hologram are allowed to reflect off of one of two separate objects before contacting the recording medium. Each object is then a potential "key" for expressing the data of the other; i.e., the resulting interference patterns recorded onto the film cannot be expressed as a visible hologram using only the reference beam: the reference beam must first be allowed to reflect off of the other object.[††] While the potential of this technique for cryptography is obvious (I would be surprised if the CIA, NSA, etc. were not already using this technique), we can also see how this would explain how a smell (or sight, taste, physical sensation, etc.)

[†] This explains so-called "photographic memory," wherein someone *always* accesses the holographic sub-quantum mind directly.

[††] For example: if you allowed one of the split-off beams to reflect off of a page of text, while the other reflected off of some small, unique object that could be easily carried on one's person—a ring, watch, necklace, etc.—the holographically-encoded page of text could not be read without the unique object being present.

can act as a "trigger" for a specific memory, as the interference data for these simultaneous sensory inputs are woven together within the holographic sub-quantum mind. And while we are fully capable of accessing memories even in the absence of all of the originally present interwoven sensory stimuli, having one or more of these stimuli present increases the vividness of the recall.

Back in the 1980s (during the cold war), the CIA developed a device that allowed them to retrieve audio data from the walls of a room up to an hour after those sounds were created.[†] When the Russians learned of this technology, they began hanging heavy tapestries on the walls of their meeting rooms, to foil the CIA's use of the device. While this may sound like high-tech science fiction, it actually utilizes an effect that is all too simple to understand. Everyone has, at one time or another, seen a demonstration of how the human voice can shatter glass. (If not, google and youtube are your friends. MythBusters did a great job of confirming this in one of its episodes.) The fundamental physical effect that allows this is known as resonance. And just as a professional singer's voice can hit the precise frequency and amplitude required to shatter a glass, the frequencies of even normal volume speaking voices make an impact upon the speaker's physical surroundings. With a sufficiently sensitive device, the history of these impacts can be read for up to an hour afterward, depending on the rigidity of the impacted surfaces.[††]

Similar technology is used by the NSA to read data from a hard drive that has been erased. By utilizing a magnetic resonance scan (or even electron microscopy), the NSA can recover data from a hard drive, even if it has been formatted several times. At last count, the "magic number" was seven; so if you re-formatted the drive and wrote over it with "gibberish" eight times, your data would truly be gone for good. But as technology improves,

---

[†] A bit like "digital psychometry."

[††] You have likely experienced this yourself: an empty room with no furniture, and nothing on the walls, echoes when you speak at normal volumes. Add some soft furniture, drapes, and paintings or tapestries on the walls, and the echo vanishes.

the NSA will also increase how many levels they can go back before data is lost to them forever.

Just as every sound makes an impact upon its environment, and every magnetic impulse is recorded on a hard drive; so, too, is every interaction of every smallest element in the universe recorded at a sub-quantum level. But in the sub-quantum infrastructure, such data is recorded across the entire infinite holographic storage medium instantly, and indelibly. At our macroscopic level, the impressions of sounds on the walls of a room fade, both due to additional sounds covering-up the previous sounds, as well as the materials in the room's walls "relaxing" back to their ground state; much as a swing will stop on its own once you stop pushing it.

Similarly, the magnetic impressions on even the most expensive hard drive will eventually fade due to slow-but-steady randomization of the magnetic material upon which those impulses were recorded. In both cases, it is *physical effects* that cause the data to eventually be lost. But at the sub-quantum level, this is not a factor. Like Bohm's "drop of ink in glycerine," all data encoded within the sub-quantum infrastructure is there for all time; an indelible record by its very nature. Our ability to glean sound from the walls of a room, or recover data from erased hard drives, is still rather primitive, as we have not been at it all that long; and our recording materials are of limited size and stability. The universal holographic sub-quantum infrastructure is truly *infinite,* and indestructible. Add to that the fact that the data is stored *holographically* (i.e., redundantly) across that entire infinite infrastructure, and we have what IT people would call a truly secure back-up. Brings the concept of "cloud storage" to an entirely new level.

To understand the actual mechanism for exactly *how* data is imprinted into the sub-quantum infrastructure, think of a ringing bell. When you ring a bell, it slowly settles down after a given amount of time. This is due to the rigidity of the material from which the bell is constructed, gravity, friction with air molecules, and internal friction within the molecules of the bell's material.

But these effects are only relevant for macroscopic objects. As we saw with the "quantum gravity problem," these forces are not a factor, even at the *quantum* level, much less the *sub*-quantum. If you picture the bell being transported to a "dimension" where gravity and friction do not exist, it would continue to ring forever, since there would be nothing to dampen the vibrations. (As Sir Isaac Newton would remind us: "An object in motion remains in motion unless acted upon by an outside force.")

This analogy is not perfect, since we are looking at the bell as a single, "solid" object. Picture that bell for what it *really* is: a holographic construct spread-out across the entire infinite holographic sub-quantum infrastructure of the universe. Considering one of the sub-quantum "pixel grid" elements to be our "bell," the "ringing," rather than being audio waveforms in a medium, is instead the pixel grid element's *data* (including state changes) being instantly propagated across the infinite holographic sub-quantum infrastructure of the universe. Similar to Bohm's ink drop being "enfolded" into the glycerine, the *data*—the *information*—conveyed by every sub-quantum "pixel" is instantly "enfolded" into every *other* pixel in the infinite universe.

Consider the previous example of the CIA retrieving audio data from the walls of a room, but only for up to one hour afterward. Why the limited time frame? Because the impressions made by the impingement of audio frequencies on those walls begins to fade rather quickly; and for the same reasons that a bell slowly stops ringing: material rigidity, gravity, and friction. But since material rigidity is a macroscopic phenomenon, and neither gravity nor friction apply at the sub-quantum level (see the "Quantum Gravity" sub-section in the "Gravity" chapter), the data can be retrieved forever without degradation: it will never cease to exist.

As I mentioned briefly above, when we utilize comparisons such as a radio sending and receiving signals in order to understand how the holographic sub-quantum mind works, we must remember that this is only an *analogy*. The holographic sub-quantum mind is not "located" at some "X-Y-Z" location "far away." Remember that, in our holographic universe, there *is* no such thing as "here" or "there"—*everything* is distributed holographically across the infinite sub-quantum infrastructure of the universe. What we perceive as the "physical world" is nothing more than an illusion.

Just as a whirlpool in a flowing river is actually nothing more than water; so, too, our holographic sub-quantum mind is nothing more than a "whirlpool" in the sub-quantum infrastructure. But so are our *bodies*. It is not a matter of our "physical bodies" being somehow in contact with a mind that is located "elsewhere;" it is that our minds and bodies are *both* holograms encoded within the infinite sub-quantum infrastructure of the universe. This "virtual reality" we call the "physical world" is simply an illusion; an experience in which we can immerse ourselves, like watching a movie, or playing a video game. You have to "suspend disbelief" in order to enjoy watching a movie, or playing a video game. You do not sit there focusing on the fact that it is nothing more than colored pixels on a flat screen, as that would prevent you from becoming immersed in the experience. What we perceive as "physical reality" is like that movie or video game: simply an illusion in which we immerse ourselves—whether for learning, or simply for fun; perhaps both. But the price of admission is the suspension of disbelief: while we are here, we have to accept the "reality" of the illusion in order to participate.

Once we understand this, we can see that, while the analogy of a radio transceiver (or a computer connecting to a remote server, web pages and "cache," cloud storage, a whirlpool in a river, a ringing bell, etc.) may help us to understand the basic concept; such analogies are not to be taken *literally*. We are all part-and-parcel of the infinite universal hologram. Any perception of physical separation is an illusion. The goal is to realize that our minds are not produced by the epiphenomena of our illusory physical bodies, as these are merely a convenient fiction that allow us to interact, for a time, with this similarly fictional environment.[†] When what we perceive as our "physical body" crumbles to dust, our holographic sub-quantum mind remains. But more on that in "Chapter 11: Life After Death."

To delve a bit more deeply into the method by which the holographic sub-quantum mind and the so-called "physical body" remain in contact, let us consider the concept of what is known as a "book code." Imagine you have a message that you want to send—securely—to someone. Once you have both agreed upon which specific book to use, all you must do is send them a page number, a sentence number, and how many letters in. You do that for each letter in the message. So, in essence, the message you want to send, *is already distributed throughout that book.*[††] But if someone does not have the key to *decipher* that message, they will not be able to retrieve it. The content of the message is *in* there, distributed in "encoded" form throughout the book; but the data is also in there for any *other* message.[†††]

In much the same way, that is how the human mind is distributed across the infinite holographic sub-quantum infrastructure of the universe. And whereas even such a large book as the Bible may have a lot of pages, and thus, a plethora of words, letters, and numbers; it is still *finite*. Thick as it may be, there are still only a finite number of pages. The holographic sub-quantum infrastructure of

---

[†] Envision our "physical world" as being like an analogy to the holographic sub-quantum mind. What is an analogy? An analogy is a descriptive linguistic device intended to illustrate a concept in a non-literal way. The "physical world" is simply an analogy; a way for the holographic sub-quantum mind to illustrate concepts to (and interact with) other holographic sub-quantum minds in a non-literal way. This entire illusory "smoke and mirrors" show we call the "physical world" is solely a method by which our holographic sub-quantum minds can interface with each other: a medium for interaction.

[††] It doesn't matter what book: you could use the Bible, it could be a cookbook, or a dictionary; as long as both you and the intended recipient of the message have access to that *exact* same book.

[†††] Because the entire alphabet is in there, and that is all you need. You are not selecting entire *words*—for instance, the phrase "tantalum capacitor" is *not* going to be in the Bible, because neither of those words existed back then. You are not selecting words; only letters and numbers. Pretty much any book you select will contain all of the numbers zero through nine, and all of the letters A through Z; and that is all you need. You may find the book lacking for punctuation in some cases—I do not know if there are many exclamation points in the Bible, for example—but you will be able to adequately convey your message.

the universe is *infinite*. So think of it like a book with an infinite number of pages.[†] This provides a truly infinite number of coordinates you could use for your "book code." Your mind—your memories, your personality— is distributed across that infinite holographic sub-quantum storage medium. So here we have another way in which it differs from a book: you could tear a page out of a book, and then when the code refers to that page, it is not there. That piece of the message is going to be missing. But with a hologram, you can cut it up into numerous pieces, and each piece still encodes the entirety. (Which is why holography is now being developed for use in computer data storage.) The data is redundantly encoded, so that any tiny piece encodes the whole. And since the universe is holographic in nature, it means you cannot "tear out a page." Thus, anywhere you go in the entire infinite universe, the data is all there. So as long as you have your "key," you can always decode the message. In the case of our book code analogy, your "key" tells you what page, what line, what letter; while in the case of our holographic sub-quantum mind, the infinite holographic sub-quantum infrastructure of the universe is our "book," but an *infinite* and *holographic* "book," so that no pages can ever be removed.

But whereas with the book code, our "key" is a piece of paper that contains a series of numbers (page number, line number, letter number) that allow you to decode the "message;" if our holographic sub-quantum mind is the "message" we seek to decode, what is our "key?" It is a complex combination of your DNA and your brain's neural pattern (and neuronal connections; but we will explore that in more depth in the "Reincarnation" chapter); because the neural pattern in your brain is the same from birth onward: you will build neural *connections* within the brain throughout your lifetime—similar to writing data onto a blank hard drive—but the neurons themselves will not change. The combination of your DNA (mainly the 95% of DNA that is designated as "junk," by the way)

[†] We will revisit this concept later in our discussion of the "Many Worlds" interpretation of quantum mechanics.

and your neural pattern—which is like a three-dimensional fingerprint in your brain that does not change over the course of your life†—will always provide a unique resonant signature that acts as your "key" to access your individual holographic sub-quantum mind; to "decode" it from the infinite quantity of other data encoded into the universal holographic sub-quantum infrastructure.

With the "book code," you could easily lose data on *either* end. If somebody tears a page out of the book, you lose the ability to decode that data. And if your "key" gets damaged so that a portion is now missing or unreadable, you will be similarly unable to decode a portion of the message. It is not that the message itself is "gone"—the data is still *there*; it simply cannot be *decoded*. With the holographic sub-quantum mind, the data is always there; and holographically distributed throughout the entire infinite sub-quantum infrastructure of the universe. If your "key" (the combination of your DNA and your neural pattern) to *access* that stored data is compromised—most likely due to physical brain damage which alters the neural pattern, or systemic DNA damage that occurs over time—the data itself is still safely stored in the holographic sub-quantum mind, but our ability to *access* the data is impaired to the extent of the physical damage to our "key." Fortunately, considering the redundant holographic nature of even our *physical* bodies (especially our brains, being the direct link to our holographic sub-quantum minds), it takes quite a lot of systemic damage before access to our holographic sub-quantum mind becomes much of a problem.

These physical analogies are necessary for us to grasp what is happening, as we are mired in the illusion of the "physical world" for our day-to-day existence. For example, you can visualize a two-dimensional square drawn on a piece of paper quite easily. You can also

---

†Unless some neurons are destroyed; which would be akin to pieces of the "key" being missing—which is what causes Alzheimer's, senile dementia, etc.—like a radio where you've damaged the circuits: clip off some resistors and a capacitors here-and-there, and that radio is not going to pull-in your favorite station so well anymore. But that does not mean that the *station* is gone: that signal is still being broadcast, even though your now-crippled radio no longer receives that signal as well as it used to.

picture that square becoming a cube in three dimensions. But try to picture how a *four*-dimensional cube would appear. It cannot be done, other than via math; because such a concept exceeds the bounds of our "physical world." Any mental image a person could conjure-up for a ***four***-dimensional cube would simply be an approximation to aid our three-dimensionally-conditioned minds in grasping what might be meant by a fourth-dimensional cube; nothing more than an abstraction.

In much the same way, the physical analogies of a radio transceiver, computer connecting to a remote server or "cloud storage," a whirlpool in a river, etc., are necessary for us to grasp the basic idea that our physical brains are neither the cause of, nor the storage location for our *minds*. Once we grasp that, we can move slightly beyond this concept, and realize that we are not talking about any kind of actual *distance* between the physical brain/body system and the holographic sub-quantum mind; rather, they are both embedded within the same infinite holographic sub-quantum infrastructure. What we perceive as the "physical world" is merely an illusion: an illusion of "things," and of separation.

Chapter 11:
Life After Death

"The boundaries which divide Life from Death are at best shadowy and vague. Who shall say where the one ends, and where the other begins?"
                                    — *Edgar Allan Poe*

I had originally intended to roll the information from this chapter into the next chapter—"Reincarnation, Mediumship & Possession"—but I felt that it was important enough to warrant its own chapter, in order to fully explore the *matter-centric* perspective from which the majority of people approach their understanding of the larger universe.

You are likely familiar with the phenomenon of ethnocentricity, wherein someone evaluates other cultures according to the standards of one's *own* culture. In polite modern society, this practice is generally frowned upon. But when it comes to such questions as whether or not there is "life after death," we are—by the very nature of the question—perpetrating this same atrocity: we are being *matter*-centric.

Throughout history, cultures world-wide have maintained widely varying beliefs concerning some form of "afterlife." But the one thing that all of these treatments of "life after death" have in common, is that they tacitly minimize whatever portion of the self that survives the death of the physical body as something that is somehow "left-over" after we die. The implication is that the body is somehow what is "important" in the equation, with whatever remains being simply a "remnant." But in reality, it is the *mind* that is most important[†]; the holographic sub-quantum mind that exists *outside* of the body. The physical body is merely a vessel— a "character" that is necessary in order to participate in this particular "video game," or *simulation* that we call the "physical world." Thus, it is not the *body* that is important, but the experiences that we accumulate while we are here in the "physical world;" and that data is stored forever within our holographic sub-quantum mind.

That said, we have to be wary of sliding into the flip-side of the matter-centric coin. In much of the esoteric literature, the "physical world" is looked down upon,

---

[†] Robert Monroe even discussed observing people attending seminars and taking classes in the "astral world." The so-called "physical world" is not the end-all, be-all that people believe it to be. It is simply one experience of *many* in which our holographic sub-quantum mind engages *simultaneously*. More on this in "Chapter 21: Astral Projection & Remote Viewing."

while "higher planes" are glorified as being somehow "better;" more "enlightened." Yes, what we call "physical world" is, in reality, noting more than a hologram—an illusion; but the other much-vaunted "planes of existence" (I detest that phrase; but it is how most people will refer to such levels as the "astral plane," the Tibetan "Bardos," etc.) are *every bit* as illusory.

In reality, *all* of the so-called "planes of existence" are illusions: holographic simulations produced by the exact same sub-quantum pixel grid; and therefore, none of these "planes" has more or less intrinsic value than any of the others. They are all *different*, so they may be useful for having different experiences; but none is any "better" or "worse" than the others. It all depends upon your perspective, and what you intend to experience by visiting the various holographic simulations.

For example: if you need to install drywall using drywall screws, an expensive "Snap-On" brand socket wrench will be about as useful to you as a soup ladle. But a common multi-bit screwdriver, while being *much* cheaper than the expensive socket wrench, will get the job done just fine. It is not that the cheap multi-bit screwdriver is inherently "better" than the socket wrench; it is simply the right tool for that particular job.

In much the same way, if you want to experience "linear time," and the mechanics of a limited "cause-and-effect-based" simulation, and the illusion of separation—whether to learn something specific, or simply for recreation—you come to the "physical world." If you want to experience something more exotic, such as the "mind instantly creating whatever the mind thinks" environment often attributed to the "astral plane," then that is where you should go. To each his or her own. Vikings would not likely enjoy living at the equator; and locals from the equator would probably not move to Iceland or Greenland. That does not mean there is something inherently wrong with either of those climates; merely that each group is uniquely suited to their own environment, and finds themselves happiest there.

But just as having the ability to travel around the world and discover what countries and climates are to your liking is a perk of living in this "physical world," so, too, does your holographic sub-quantum mind have the ability to sample various other "worlds." (Of which, in a truly infinite universe, there is—by definition— an infinite variety.) Thus, when we hear someone ask if there is "life after death," we can examine this question from the "big picture" viewpoint, and realize that one's holographic sub-quantum mind can be experiencing many different "planes of reality" simultaneously; all of them just as illusory as the next; but with each offering us something unique and interesting enough to have drawn our attention in the first place. After all, would you be happy with a TV that had only one channel?

Once we can get beyond the "matter-centric" mindset, we can see that which "plane of existence" we currently believe ourselves to inhabit is simply a matter of where we focus our attention; like tuning into a specific radio station for a moment. Alternate "planes of existence" are similar to the "bleed over" from adjacent radio frequencies when you are roughly equidistant from two stations broadcasting on similar frequencies. Or if you are listening to an older radio with poor frequency discrimination.

While existing within the "physical world," we are "tuned" to matter; but while we are asleep (or in other altered states of consciousness), we can "drift" partially, temporarily, into adjacent "frequencies." (More on this in "Chapter 21: Astral Projection & Remote Viewing.")

Thinking of the universe as an infinite "super computer" running an infinite number of "simulations," the three-dimensional "physical world" in which *we* are currently mired is but the tiniest infinitesimal portion of the vast super computer's capabilities. This is why maintaining a matter-centric perspective makes absolutely *no* sense. We have to open our minds to the "big picture," and realize that what we perceive as the "physical world" is only the smallest inconsequential slice of the overall pie. Much as our eyes are only capable of perceiving the slimmest portion of the electromagnetic spectrum—that portion we refer to as "visible light"— our minds (at least from our current matter-bound perspective) are only capable of directly perceiving a *very* small portion of the infinite universe. If this slim portion of the universe that we can directly perceive were truly "all that is," it would be an extreme waste of resources, tantamount to having the world's most powerful supercomputer at our disposal, and using it only to play "Pong."

The other "programs" being run on the infinite sub-quantum supercomputer are such things as the "astral worlds," the Tibetan "Bardos," and a host of others, *ad infinitum*. Just as a TV can display a near infinite variety of programs, all of the potential for an infinite number of other such "planes of reality" (as *we* call them, from our matter-centric perspective) is inherent within the sub-quantum pixel grid. We will explore this concept more deeply in "Chapter 19: The 'Many Worlds' Interpretation Of Quantum Mechanics." For now, the important aspect to grasp is the "matter-centric" concept; so we can get *beyond* it. We need to expand our minds in order to apprehend the "big picture." It all comes down to *perspective.*

The mind *does*, indeed, survive the death of the "physical" body, as your mind is encoded within an infinite universal hologram, and can never be destroyed. All of your memories and personality—the entire essence of your mind—are preserved for all time. This will eventually allow us to access the minds of "dead" people (once we develop the proper technology; e.g., a sub-quantum computer that can be "tuned" to any holographic sub-quantum signature) in much the same way that natural "mediums" do. Just think: we could boot-up a sub-quantum computer, and connect to the mind of *Tesla!*

And how does this affect the notion of "life after death"—i.e., the idea that the mind somehow continues "without a body" in some "other world?" You must remember that what we call our "physical bodies" are nothing of the sort. As we discussed previously, even a mainstream physicist will tell you that we are mostly "empty space"—that "force fields" are responsible for objects feeling "solid" to us; similar to magnets exerting force against *each other*, while *we* feel nothing "solid" between them.

So, if our "physical bodies" are not even "real"—objectively speaking, from the standpoint of the universe at-large—then why should their expiration matter? We actually exist simultaneously across many "bodies" on a multitude of different "planes." There is no logical reason to assume the primacy of what we perceive as "physical bodies," relegating all other "less tangible bodies" (from our matter-centric perspective) as being somehow "slaved" to *this* one. In fact, astral projectors like Robert Monroe report that, in what they call the "Astral World," objects feel as solid to them when they are there, as objects *here* feel to our "physical" bodies. People on the "astral" level can *pass right through* "solid objects" that exist on *this* level; so from *their* perspective, *we* are less tangible—or less "real"—than *they* are. It is all a matter of perspective. Which is to say, that simply because two different objects are in different *phases* does not make either one more "real" than the other. Thus, to consider one "plane of existence" to be more "real" than the other is simply a matter of illogical prejudice in favor of one's own "home plane." There's that *matter-centric* thing rearing its ugly head again.

The question "Is there life after death?" is a matter-centric question; and one that, by its very asking, forces one to adopt a limited, biased, matter-centric perspective.

Since we already understand the "pixel grid" analogy, let us extend that, but in a different way. If we think of the infinite sub-quantum pixel grid as being similar to a TV, and "physical matter" as one *program* that is regularly viewed on that TV; what if that particular "program" is *canceled?* Does the TV cease to exist? Of course not!

The "physical world" experience is simply one "program" that is being displayed by the infinite sub-quantum pixel grid. So when the show is "over" for us (i.e., after our body's physical death), we are still conscious—and therefore, still connected to and interacting with our unique, individual holographic sub-quantum mind—on an infinite number of *other* "channels." We may or may not choose to take part in *this* "channel"—the one we call the "physical world"—some time again in the future; but whether we do so or not is irrelevant: our holographic sub-quantum mind will remain active, and continue to have experiences on an unfathomable number of other "channels."

The so-called "physical body" is, in actuality, no such thing—*all* matter is nothing but energy; and even *energy* is, at its most fundamental level, truly "nothing at all." It is all an illusion. So when the fictional "physical body" ceases to exist, what does it really matter? To our holographic sub-quantum mind and our larger self, in the grand scheme of things, it makes very little difference.

Chapter 12:
Reincarnation, Mediumship
& Possession

"It is by no means an irrational fancy that, in a future existence, we shall look upon what we think our present existence, as a dream."
— *Edgar Allan Poe*

Once we realize, as discussed in the previous chapter, that what is commonly referred to as "life after death" is real (though not exactly as commonly envisioned), the next question that comes to mind is "What about reincarnation; is it real as well?" And again, the answer is a qualified "yes." But as with "life after death," not in the manner generally portrayed.

What is commonly known as "reincarnation" is, in fact, a real phenomenon; but it does not function in exactly the way most people believe, and most certainly not as portrayed by Hollywood and popular "New Age" books.

Russian psychiatrist Vladimir Raikov, M.D. developed a technique he dubbed "artificial reincarnation," wherein experienced hypnotic subjects capable of reaching the deepest levels of hypnosis were instructed to "become" famous painters and musicians. University students were polled, and subjects were chosen specifically for their complete lack of previous talent in the areas of art and music. But during the "artificial reincarnation" experiments, the students were able to paint like the great masters, and play musical instruments as if they were Mozart or Rachmaninov.

People who *naturally* possess the ability to temporarily "re-tune" into holographic sub-quantum minds other than their own are referred to as "mediums" or "channelers." Since Raikov was able to duplicate this ability at will, temporarily enabling virtually *anyone* to "become" Mozart, Rembrandt, etc., this demonstrates that the person is not actually the "reincarnation" of that deceased individual; they are simply able to access the data bank— the holographic sub-quantum mind—of that person.

The memories stored within your holographic sub-quantum mind are like data stored on a hard drive (think "Microsoft Word" files, for example), while the personality (thought patterns, emotional state, etc.) are like *programs*. Some mediums (and some hypnotic regressions) can only access the holographic sub-quantum mind of a deceased person to a limited extent; basically, just the *data*. More adept mediums (and more advanced hypnotic subjects) can access (i.e., actually *"run"*) the *programs*—the *personality* of the deceased individual. This ability spans a range that, at its uppermost extreme, encompasses the state generally referred to as "possession," wherein a medium becomes "locked-into" the running programs that they have accessed from someone else's holographic sub-quantum mind.

This explains why sometimes you can think back and remember *what* happened during an event from your past—you can still access the *data* from a past memory—but you can no longer "get inside the head" of the person that you were then.[†] Hypnotic regression can temporarily allow you access to the "programs" from that younger version of yourself; with the side effect that, for the duration of the regression, you are "stuck" within the mental state of that previous version of yourself. This is why many hypnotists prefer to have you remain in your present state of mind, and simply access the *data* from a previous time in your life, instructing you to view it "on a screen, as if watching TV, or a movie at the theater." It is often less traumatic for people to relive difficult scenes from their earlier life if done in this fashion. But for therapeutic value, some regressive hypnotists will *intentionally* have the person fully "re-live" the event, in order to resolve deep-seated issues. This technique is similar to Raikov's "artificial reincarnation," but instead of temporarily becoming someone *else*, you are becoming an earlier version of *yourself*.

In regression hypnosis, people who are hypnotically regressed to younger periods in their life actually change their speech patterns, affect, and body language while tuning-into the thoughts and memories from that earlier time. By using hypnosis to regress a person to a younger version of themselves, they can then resonate with

---

[†] Think of trying to open old Photoshop documents on a newer computer. The newest version of Photoshop can easily open old .PSD files. This is known as "backward compatibility." But try to install and run the old version of Photoshop *itself* on one of the newest, latest, greatest computers. Without installing some sort of emulator, you will not be able to do that, as the operating system *and* the hardware have changed too much to run that ancient version of Photoshop. But whereas we humans are stuck with our current hardware and operating system, the mind is *much* more plastic, and can relatively easily revert to a previous state; almost as if Apple's "Time Machine" back-up software were installed in our minds.

the thoughts and memories of themselves at that age. For all intents and purposes, for the length of the session, they *are* that younger version of themselves.[†]

There are also a plethora of reports concerning reincarnation among people who have *not* been hypnotized. And though these individuals have not been "suggested" into re-tuning to the holographic sub-quantum mind of a deceased individual, the underlying phenomenon is the same. Your body and brain are simply a very complex "tuning" mechanism that allow you to access *your* unique, individual holographic sub-quantum mind. But since all human DNA is very similar, it occasionally happens that there is some partial overlap, just as when a cordless phone accidentally tunes-in to someone else's phone's frequency. The early cordless phones used to experience "cross-over"—where you would hear someone else's conversation on your phone. This occurred because there were only so many frequencies allotted by the FCC to cordless phones; so when two people who had phones on the same (or even very *close*) frequencies lived close enough to one another, you would get the "party line" effect. The same effect can occur—to varying degrees— with the holographic sub-quantum mind. In much the same way that early cordless phones could interfere with each other due to a limited frequency spectrum; so, too, can the unique holographic sub-quantum signatures of two *people* partially overlap. This is generally known as "reincarnation"—when a person is born whose holographic sub-quantum signature overlaps, to some extent, with the holographic sub-quantum signature of a deceased individual. But "mediumship" is a result of the exact same phenomenon, with the difference that a medium has the ability to "re-tune" at will (usually by utilizing some object that has been "imprinted" with the deceased person's holographic sub-quantum signature)

[†]Think again of Apple's "Time Machine" back-up software, where you can "rewind" your computer to a specific date in the past, in order to retrieve data that you accidentally deleted. It is very much like that: you are resonating with your holographic sub-quantum mind as it was at that earlier time; so it is as if your holographic sub-quantum mind has been "rewound" to that exact state. In a way, it is almost like a medium "tuning-in" to a deceased person's holographic sub-quantum mind; except that it is your *own*.

into the holographic sub-quantum minds of others. A medium can access (to varying degrees, depending upon the extent of their ability) the memories, thoughts, and even—in extreme cases—the *personality* of the deceased person. We can thus see that what is generally known as "reincarnation" is merely a more permanent version of what any adept medium can accomplish.

If a person becomes very attuned to an historical figure, they can also "tune-in"—to varying degrees—to that person's holographic sub-quantum mind; sometimes just for bits and pieces of data; and sometimes, for much more. Mediums can accomplish this fairly easily; but they do so intentionally—like a self-directed version of Raikov's "artificial reincarnation."

In combination with Raikov's "artificial reincarnation," realizing that the sub-quantum infrastructure of the universe (being an infinite holographic storage medium) is what the ancient Tibetans called the "Akashic Records"—the data bank of all that is; what Carl Jung called the "Collective Unconscious"—not only explains what is *commonly* known as "reincarnation," but also why many different people, living simultaneously, claim to be the "reincarnation" of some historical figure: they are all tuning-in to that person's holographic sub-quantum mind.[†] In simple cases where a person is only "tuned-in" enough to receive data, they can "remember" the facts and events of the deceased person's life, as if they were their *own* memories. In cases of mediumship, they are tuned-in to a *greater* extent, allowing them to access, not only the *data* (memories), but to actually activate the "programs," or *personality* contained within the holographic sub-quantum mind, at least to a limited extent. In *extreme* cases, this can lead to what is commonly known as "possession," where the person becomes so tuned-in to the holographic sub-quantum mind of another, that the personality

---

[†] This has always been one of the main arguments against the traditional beliefs in reincarnation and "karma." If a person experiences serial lifetimes, accumulating "karmic debt" in one lifetime that must be paid for in the next; how could multiple individuals *all* be the reincarnation of someone who is dead? This paradox is resolved when we realize the truth about how "reincarnation" *really* functions.

they are accessing takes-over, temporarily displacing that person's connection to their *own* holographic sub-quantum mind.

Possession is, therefore, the extreme end of this "tuning-in" phenomenon, wherein the "programs" that comprise the deceased person's personality temporarily take over the body of a currently living person. People who are prone to possession are similar to naturally gifted mediums, but without the ability to control the phenomenon. Picture as it being like a person afflicted with Tourette's Syndrome: it *is,* in a manner of speaking, a "defect"—like cars that are "lemons," or electronic devices that behave in a "quirky" manner. But whatever might be wrong with such a person's "circuitry," they have the ability—like a gifted medium—to tune-in to holographic sub-quantum minds other than their *own*, and to do so quite *intensely*. But what they lack is the ability to *control* this faculty. That is the defect. But just as a faulty electronic device or vehicle can generally be repaired, people who are prone to possession may likewise learn to control their abilities. Breathing, meditation, hypnosis, dietary changes; and most of all, *practice,* may enable them to transition from being victims of "possession," to being gifted mediums in full control of their faculties.

Possession generally occurs when a susceptible person with certain natural abilities (i.e., extreme mediumship, often combined with strong psychometric abilities) either becomes obsessed with an historical figure, or (via their psychometric ability) even simply comes into contact with an object that has been strongly imprinted with a deceased person's unique holographic sub-quantum imprint.† (More on this in "Chapter 24: Ghosts & Hauntings.") This causes the person to resonate so strongly with the deceased person's holographic sub-quantum mind that, in effect, they temporarily "become" that person. For all intents and purposes, it is as if the deceased person's holographic sub-quantum mind once again has a physical body;

---

†This psychometric aspect may be the root of the myth of "cursed" objects, and why they affect certain people, but not others.

at least for as long as the "possession" lasts. The severity and duration of the possession depends upon the extent of the resonance. It is usually temporary, as physical fatigue tends to spontaneously "de-tune" the possessed individual, allowing them to return to their default state. But if the source of this phenomenon is not understood, the condition can recur at any time. Once resonance with a holographic sub-quantum mind is established, it is easier to re-establish the connection in the future. This is where education and training can help the victim of possession to control their ability, and prevent future unpleasant incidents.

I must be clear, and reiterate that it is only *data* that has been stored on the "hard drive" that is the holographic sub-quantum mind. But that is *all* that is saved: the data; the essence of who and what a person is; the sum total of a person's experiences. And if the holographic sub-quantum mind of a deceased person is accessed by a medium, that data can be retrieved. If someone is in strong resonance with the holographic sub-quantum mind of a deceased person for an extended period, new data (thoughts and memories) can even be *added* to that person's sub-quantum "hard drive." In the previously-discussed case of someone being born whose holographic sub-quantum signature *overlaps* with that of a deceased person, they might *think* of that as being the popularized notion of a "reincarnation;" but it is not. It is simply somebody "tuning-into" the deceased person's data storage. But since the link is strong, new data (thoughts and memories) *can* be added to the deceased person's holographic sub-quantum mind; just as someone who obtains your computer or e-mail password can access your current files, as well as add new data.

That said, we must remember that the past physical existence(s) of the owner of the holographic sub-quantum mind that is being accessed represented merely the barest sliver of that mind's infinite spectrum of consciousness. Since they remain conscious across the *remainder* of that spectrum following the death of their physical body (remember our "matter-centric" discussion in the previous chapter), someone in the *present* physical world accessing their holographic sub-quantum mind in such a limited way (and for such a limited period of time) is of no more consequence than someone playing your character (i.e., using your "save file") in one of your favorite video games while you are away from home. Perhaps even *less* significant, for whereas someone playing your video game character could conceivably *lose* money or items that your character possesses within the game, or somehow cost you experience points; someone temporarily accessing a deceased person's holographic sub-quantum mind can only *add* to the store of experiences contained therein.

I say this to help avoid the tendency to lapse into a "new age/our 'soul' survives death" tangent. Such a "spiritual" interpretation of reincarnation is a pitfall to be avoided, as it is not supported by the facts. In order to remain scientific about this concept, we must steer clear of such unscientific notions, as they destroy objectivity. And once you lose objectivity, you no longer have a science.

The concept of "karma"—as discussed in the footnote on page 130—is another spiritual belief with no basis in fact. In order for the spiritual notion of "karma" to operate as specified in various religious belief systems, reincarnation must also operate according to the tenets of those belief systems. Since we have seen that reincarnation does *not*, in fact, operate according to those tenets, the fantasy known as "karma" is null-and-void. In the *real* world, we do not get what we "deserve," but rather, what we *expect*. (More on that in the "Manifesting" sub-section of "Chapter 22: Magic(k).")

Chapter 13:
Genetic Memory

"It is high time public health researchers took human transgenerational responses seriously."
— *Professor Marcus Pembrey, Geneticist*
University College, London

While reading the previous chapter, I am certain it occurred to many an astute reader that, if DNA is instrumental in tuning-in to one's holographic sub-quantum mind; and further, if even *slight* "overlap" of holographic sub-quantum signatures can lead to what is generally known as "reincarnation," then might not the phenomenon of "genetic memory" be attributed to the similarity of DNA shared by relatives?

Indeed it is. In fact, I consider it such a foregone conclusion, that I nearly relegated the idea to a mere footnote in the previous chapter. But upon further reflection, I decided the concept is important enough to warrant at least a page or two of its own; especially in light of the acceptance the concept of "genetic memory" is finally gaining in mainstream circles.[†] Granted, the mainstream is—as usual—barking up the wrong tree in thinking that the memory data is literally stored *within* the physical DNA; but it is impressive nonetheless that they have at least advanced from considering the concept of genetic memory to be "pseudo-science" and "fringe nonsense," to accepting it as a valid phenomenon. In light of the amount of evidence for the phenomenon, they could hardy have done otherwise; it was simply a matter of time. That said, it must be noted that they are on the wrong track, and would be better served to consider the "encryption key" aspect of DNA.

We can readily apprehend that, with DNA being so instrumental in the "tuning" process, the similarity of DNA between people of close genetic relation allows offspring to "tune-in" to portions of their ancestors' holographic sub-quantum minds. This may happen from birth, as occurs with many cases of "reincarnation;" but the phenomenon can also present itself later in life, when an interest in one's ancestors allows one to more actively establish a resonance with the holographic sub-quantum minds of the deceased, much as occurs with mediumship or possession. This may explain the common

---

† http://tinyurl.com/pnjugw3
  (If this link ever ceases to function, you can download a web archive .zip file at the following URL: http://tinyurl.com/pdjwed6)

practice among many cultures of ancestor worship. After all, with shared DNA facilitating the process, it should be much easier to "contact" (i.e., resonate with the holographic sub-quantum mind of) a deceased relative than a stranger. This would have indicated to shamanic cultures in millennia past that there was a stronger "link" between the living and their deceased ancestors than there would be with deceased *non*-relatives; and they were quite correct in this belief. The only aspect that has changed in our consideration of genetic memory in the modern era is that now, we can understand the actual *science* behind this phenomenon.

It is, in fact, our modern understanding of the underlying science behind the phenomenon of genetic memory that allows us to speculate that perhaps Raikov's "artificial reincarnation" procedure would be easier to accomplish, and have longer-lasting effects, if one were to choose an ancestor of the hypnotic subject as the source of ability—i.e., an ancestor possessed during their lifetime of some special skill that the subject wishes to acquire—rather than "tuning-in" to the likes of Rembrandt or Mozart. I would posit that, not only would the procedure be easier to accomplish under such conditions, but that the effects may be longer-lasting upon the conclusion of the hypnotic session than when non-relatives such as Rembrandt or Mozart are the source of the talent.

But while the stronger innate resonance with a deceased relative (though said relative needn't necessarily be deceased—more on this in "Chapter 15: Telepathy") may facilitate the "artificial reincarnation" process, and produce lasting results; there may also be a higher risk of possession. Researchers in this area must be aware of this potential, and prepare subjects accordingly. And I do believe we will see much research in this area in the coming years.

The same fundamental phenomenon responsible for reincarnation, mediumship, possession, and genetic memory also underlies such mental dissociative disorders as MPD ("Multiple Personality Disorder," now more commonly known as "DID," or "Dissociative Identity Disorder") and Schizophrenia. These disorders cause the victim to "tune-in" to multiple holographic sub-quantum minds at random, with little-to-no ability to control when this occurs; similar to possession. All of this switching back-and-forth between holographic sub-quantum minds by someone who has had no training in how to deal with such a phenomenon (as a medium would have) can lead to that person having a tenuous link even with their *own* holographic sub-quantum mind—they often end-up "seeing through the cracks" in our "consensus reality;" and therefore, they exhibit the characteristics commonly associated with dissociative disorders.

A similar disorder related to the inability to connect to one's own holographic sub-quantum mind is at work in a coma patient. Rather than connecting to various holographic sub-quantum minds in an uncontrolled manner, as happens with MPD/DID and Schizophrenia (or to a more severe level, as with possession), a coma patient loses the ability to link to *any* holographic sub-quantum mind, including their own. Sometimes, the damage that caused this total "disconnect" is eventually able to repair itself (since cells are still dividing, wounds are healing, etc.). And sometimes, it cannot, and they die before ever emerging from the coma.

I find myself wondering if, once the concept of a holographic sub-quantum mind is mainstream, and medical personnel/psychiatrists are trained to deal with issues on a sub-quantum level; will a competent mental health practitioner be able to make contact (as would a medium) with the holographic sub-quantum mind of the coma victim on one of the *other* levels on which they *are* still conscious, and say "Um, you *do* realize that you still have a living, breathing body on the 'physical world' level, right? You seem to have left it unattended. It would be very thoughtful of you if you would please re-animate it, as we could really use the bed space and resources for *other* patients who actually *need* our help!"

Chapter 14:
Hive Minds/Group Minds

"A more or less superficial layer of the unconscious is undoubtedly personal. I call it the 'personal unconscious.' But this personal layer rests upon a deeper layer, which does not derive from personal experience and is not a personal acquisition but is inborn. This deeper layer I call the 'collective unconscious.' I have chosen the term 'collective' because this part of the unconscious is not individual but universal; in contrast to the personal psyche, it has contents and modes of behaviour that are more or less the same everywhere and in all individuals."
— *Carl Jung,*
"The Archetypes and the Collective Unconscious" (1934)

In the preceding chapters, we have thus far addressed only *individual* holographic sub-quantum minds; but we need to briefly discuss the concept of hive minds and group minds, as this concept will become quite important in subsequent chapters.

Every living thing has a holographic sub-quantum mind; but as the neurological infrastructures of living creatures decrease in complexity, their holographic sub-quantum minds become increasingly less individual, due to the greater degree of overlap between members of that species. The neurological structures of insects are not complex enough to warrant individual holographic sub-quantum minds: they only have identity as a *hive mind*.

In more complex creatures possessing individual personalities, there exists an individual holographic sub-quantum mind, as well as a *group* holographic sub-quantum mind that encompasses (and is shared by) the entire species. This *group* holographic sub-quantum mind is what Carl Jung referred to as the "Collective Unconscious." (It was also known as the "Akashic Records" in ancient Tibet. More on that in chapter eighteen.) This *group* holographic sub-quantum mind is, by its very nature, more of a "catch-all"—i.e., less specific, more archetypal—a bit of a "jack of all trades, master of none." This is why mobs never behave as rationally as the individual members of said mob might do on an individual basis. Once individuals give themselves over to mob behavior, they are all resonating with an aspect— a *subset*—of the *group* holographic sub-quantum mind, and have momentarily surrendered the will of their own *individual* mind to the group's will. This is why a person who can exist in the normal day-to-day world as a scientist, doctor, lawyer, banker, etc., often thinks and behaves quite differently while acting as part of a *mob*.

To clarify the concept of *subset* group minds within the overall *species* group holographic sub-quantum mind; consider, as an example, devout Catholics: they believe in such "miraculous" acts as transubstantiation, stigmata, and bleeding/crying statues. While Catholics are still members of the *overall* group holographic sub-quantum mind of the human species as a whole, they *also* have their *own* group holographic sub-quantum mind that is a *subset* of the larger group mind. This is why the Catholics experience such "miracles" as stigmata, while the Protestants do not. (We will discuss this further in "Chapter 22: Magic(k).") The rioting mob mentioned a moment ago is another example of a localized *subset* group holographic sub-quantum mind.

In some instances—such as within mind-control cults—the *individual* holographic sub-quantum mind and its attendant personality is *intentionally* suppressed to a large extent, limiting the individual to functioning only as part of the *subset* group holographic sub-quantum mind of that particular cult. In this instance, the group mind of the cult is behaving much the same as the hive mind of less complex creatures. But this situation is not healthy, and—much like possession—cannot be maintained indefinitely.

The more *positive* aspects of the group mind phenomenon have the potential to alter the future evolution of mankind. After all, super-computers are created by assembling massively parallel arrays of individual processors; and the same process can allow individual minds to link together for a common undertaking. (More on this in the "Maharishi Effect" subsection of chapter 22.) We can only hope that, as a higher percentage of the world's population evolves to possess greater control over the connection to their own *individual* holographic sub-quantum minds, we see a drastic reduction in mob behavior, and an increase in positive "parallel processing" undertakings.

Chapter 15:
Telepathy

"In science fiction, telepaths often communicate across language barriers, since thoughts are considered to be universal. However, this might not be true. Emotions and feelings may well be nonverbal and universal, so that one could telepathically send them to anyone, but rational thinking is so closely tied to language that it is very unlikely that complex thoughts could be sent across language barriers. Words will still be sent telepathically in their original language."
— Michio Kaku, The Future of the Mind: The Scientific Quest to Understand, Enhance, and Empower the Mind

We are not even going to ask the question, "Is the phenomenon of telepathy real?" as there is ample evidence, going back many decades, that it *is*, indeed, a real phenomenon. The most recent efforts along these lines have been conducted by Dr. Dean Radin, as detailed in his latest book "Supernormal." (Highly recommended reading for those of you who may still have doubts as to the validity of the phenomenon of telepathy.) Having thus moved *past* the "Is it real?" stage, we can move on to exactly *how* the phenomenon works.

Self-described "skeptics" (though the vast majority of them are actually "debunkers") have always heaped scorn upon the idea of telepathy, chiefly because proponents of telepathy tend to claim that it is the *physical brain* that somehow "sends and receives signals—like radio waves." Skeptics correctly point out that the electrical signals produced by physical brains are *far* too weak, and would scarcely penetrate the skull of the sender, much less propagate across any significant distance to be received by another physical brain that was attempting to "tune-in" to the transmission.

That is, in fact, not at *all* how telepathy actually works. But this is where the self-described "skeptics" prove themselves to be merely "debunkers." A true skeptic is a true scientist, in that they remain unconvinced that a phenomenon is real until the facts *prove* that it is real. Thus, while a *true* skeptic would initially doubt that telepathy was real; upon reviewing the evidence (including the recent controlled laboratory tests conducted by Dean Radin et al), such a skeptic would be convinced of the reality of the phenomenon of telepathy. That said, such a true skeptic would also quickly shoot-down the above-mentioned notion that the physical brain sends out "radio waves" in order to accomplish telepathy. But shooting down someone's flawed theory concerning exactly *how* a phenomenon functions does not invalidate the phenomenon *itself*. After all, as we discussed in "Chapter 4: Gravity," any honest physicist will readily admit that they have *no idea* how gravity actually works. A few hackneyed theories aside, modern mainstream science simply has no real idea what causes gravity. But that does not mean that gravity does not *exist*. The evidence for its reality is all around us. So in spite of the fact that no current theory of gravity within the scientific mainstream has proved correct, the phenomenon itself remains quite real. And this is where the self-described "skeptics" who disdain telepathy based upon the faulty explanation advanced by some *proponents* of telepathy prove themselves to be "debunkers" rather than *true* skeptics. A true skeptic would realize—based upon the data—that the *phenomenon* of telepathy is real; but at the same time, would happily explain to proponents of the "brain emits radio waves" theory that said theory was wrong. A true skeptic would not "throw the baby out with the bath water" by concluding that, because some proponents of telepathy advanced a faulty theory to explain *how* telepathy works, the phenomenon itself is not real. Such a position would be illogical, and no true skeptic would fall prey to such behavior.

Realizing that the phenomenon of telepathy *does*, in fact, exist, we can easily understand its mode of operation via the holographic sub-quantum mind. Since the mind of every living creature is distributed across the infinite holographic sub-quantum infrastructure of the universe (every smallest portion of which communicates with every *other* smallest portion *instantaneously*); there is, in effect, *zero distance* between the holographic sub-quantum minds of any two individuals. Thus, by one holographic sub-quantum mind being even *partially* in resonance with another, this facilitates the exchange of information—thoughts, feelings, memories; and in some cases, even personality traits—between those two minds. This operates upon the exact same principals as the previously discussed phenomena of reincarnation, mediumship, and genetic memory. Regardless of *how* two holographic sub-quantum minds come to be in some degree of resonance with each other, they are thereafter able to exchange information to a degree limited only by the *extent* of that resonance.

In chapter thirteen, I mentioned that the phenomenon of genetic memory need not necessarily involve a *deceased* relative. If we reflect upon the fact that the DNA of twins is identical (or very nearly so, barring copy number variants; but that is splitting hairs), we can see how having such a major component of the "encryption key" (which allows one's physical body to connect with one's individual holographic sub-quantum mind) in common with another individual, would lead to a high degree of resonance between those two individuals. This is precisely why we see such a high incidence of telepathy between twins as compared to the rest of the population.[†] It also explains why mothers tend to "just know" when something has happened to one of their children, though they may be a continent away.

Telepathy occurs much more frequently than most people realize, as it is usually written off as "coincidence." But at least a certain percentage of people who "hear voices" may, in fact, be unintentionally "tuning-in" to the holographic sub-quantum minds of one or more other people (living or deceased); or even several holographic sub-quantum minds *simultaneously,* at *random,* leading to a diagnosis of Schizophrenia, MPD/DID, etc., as we discussed in chapter 13. Imagine how difficult it would be to function normally in society if this were happening to you; and not only could you not comprehend what was happening, but you were also unable to "turn it off."

To wrap-up on a lighter note than *that* grim thought; a question I often receive at seminars and via e-mail from pet-lovers is: "Does telepathy also exist between people and animals?" Yes, indeed; a person can absolutely resonate with the holographic sub-quantum mind of a pet.

---

†Since it is a demonstrable *fact* that twins exhibit a higher incidence of telepathy—both with regard to frequency of events as well as the average amount of information exchanged; i.e., the extent to which they are "tuned-in" to one another—it is quite telling that, while the prevailing notion among proponents of telepathy that the physical brain somehow transmits "radio-like waves" from one physical brain to another does not hold-up under scrutiny; we can see that, not only does the holographic sub-quantum mind explanation for telepathy explain how telepathy in *general* works; but it would also fully *expect* that twins (as well anyone sharing DNA, such as family members) would experience this phenomenon to a greater degree than would people who are not related.

They will not be able to "converse," per se, since they do not share a language in common; but communication of a basic nature can occur. This phenomenon is quite common in shamanic cultures; it is simply foreign in our Western culture. But speaking globally, we are the exception in that case rather than the rule.

Chapter 16:
Immortality

"I would love to believe that when I die I will live again, that some thinking, feeling, remembering part of me will continue. But as much as I want to believe that, and despite the ancient and worldwide cultural traditions that assert an afterlife, I know of nothing to suggest that it is more than wishful thinking."
                                        — Carl Sagan

Since our holographic sub-quantum mind is encoded into the infinite sub-quantum infrastructure of the universe; as far as true immortality is concerned, that would be impossible to top. But from our *matter-centric* perspective, when people say "immortality," what they really mean is the extended continued existence of their current ego consciousness in a *body* of some sort, here in our physical reality. Given *that* definition of immortality, unless you wish to include Ponce de León's mythical "Fountain of Youth," the only real contender (hypothetical though it currently remains) is what is commonly referred to as "uploading." This proposed method of achieving physical immortality involves "somehow" uploading a person's consciousness into a computer, and it is a ridiculous notion for two reasons: first and foremost, since your mind does not exist within the confines of the physical brain, there is nothing to upload. Your holographic sub-quantum mind is distributed across the infinite sub-quantum infrastructure of the entire universe. Thus, it *already exists* at every point in the infinite universe; so the idea of "uploading" it from one place to another is nonsensical. There is absolutely no need for it to be "transferred" from your physical brain into some other physical device.

In order to utilize computer technology to achieve physical immortality, the computer would have to be capable of achieving 100% resonance with your holographic sub-quantum mind; meaning you would have to build a computer that could duplicate your holographic sub-quantum signature—your "encryption key"—perfectly. And that brings us to point two: in order to accomplish that, you must have a computer capable of interacting with the sub-quantum infrastructure of the universe; and we have not, as yet, even perfected *quantum* computers, much less *sub*-quantum computers.

So, will it one day be possible to achieve physical immortality via computer technology? Absolutely. Just not by utilizing the notion of "uploading" which is currently so popular. If proponents of the "uploading" method continue to think of it in that manner, they will never achieve success. But if they were to change their focus, and seek instead to produce a computer that is capable of interacting with the universe's sub-quantum infrastructure; once said computer is capable of perfectly resonating with the subject's holographic sub-quantum mind (or the holographic sub-quantum mind of a deceased relative or historical figure), they will eventually be able to achieve their goal of physical immortality via computer technology.

That said, there will be the issue of how to establish resonance with the subject's specific holographic sub-quantum mind. The computer would have to be capable of performing a sub-quantum analysis—either directly upon the subject, or of some object that has been strongly imprinted with the subject's holographic sub-quantum signature, similar to the personal items a medium uses to contact the holographic sub-quantum mind of a deceased person—to determine the proper holographic sub-quantum signature to use. In essence, the sub-quantum computer would be replicating the abilities of a medium, albeit in a more technological manner. But in the absence of either the person themselves (if they are still among the living), or an object strongly imprinted with their holographic sub-quantum signature, the chance of success would be remote. This would be the case in most instances where you were trying to tune the sub-quantum computer to the resonant signature of a deceased relative or historical figure, as there would be no sufficiently imprinted object to assist with the tuning process. It would be the equivalent of a ham radio operator simply continuing to turn the tuning dial until they connected to a strong signal; and then you would have to ascertain the identity of the holographic sub-quantum mind you have contacted. Finding a specific needle in an entire *pile* of identical needles would be simple by comparison. Unless you have an imprinted object to assist with tuning-in to a specific holographic sub-quantum mind, any effort to connect with the holographic sub-quantum mind of a particular person is almost certain to fail. For this reason, once the technology is perfected and relatively well-known, I can envision it becoming commonplace for people to preserve at least some small piece of a deceased relative, in hopes of one day utilizing it to tune a sub-quantum computer (they are certain to be expensive, at least for some time after their initial creation) to the holographic sub-quantum mind of the deceased.[†]

We must bear in mind that, as we discussed in previous chapters, the "physical world" comprises only the barest sliver of the overall infinite spectrum of consciousness; and your holographic sub-quantum mind is simultaneously conscious across a multitude of other levels within that spectrum. In the larger scheme of things, the "physical world" portion of the overall spectrum is relatively inconsequential. Effectively, you already *are* immortal; but for people who are wanting to hold onto their current ego consciousness— looking at it from the traditional "reincarnation" standpoint: their current "incarnation"—they do not want to "shuffle off this mortal coil;" they want to maintain their presence on this level, with their current personality intact. Thus, you have to consider whether someone of that mindset would actually want to be trapped inside of a large (as the first sub-quantum computers will most likely be quite large, as are all first generation computers of a fundamentally new type)— and therefore, stationary—computer; at least until a sub-quantum computer small enough to fit within an ambulatory robot is perfected. These ambulatory robots might eventually resemble something less like Japan's current "Asimo" robot, and more akin to those depicted in the movie "Surrogates." With this "sub-quantum computer built-into a robot body" paradigm, even if the robot body were ever destroyed, another one could simply be

---

[†]This would also be useful in facilitating the process to be discussed in "Chapter 17: Holographic Cloning."

tuned to the proper holographic sub-quantum signature.[†]
In effect, this would be true physical immortality.
You would simply need a way to *store* your holographic
sub-quantum signature in some way; just as music,
movies, etc. are now stored in data files. We will discuss
that further in the next chapter.

[†] This begs the question of whether or not one could have *multiple* robots all tuned-in
to the same holographic sub-quantum mind, thereby producing a "multiple parallel
incarnation" effect. In theory, this should be possible; though whether or not the human
mind could easily adapt to such a potentially disorienting state of affairs will remain
to be seen once testing becomes possible. My money is on the innate adaptability of
humans being able to manage it quite nicely.

Chapter 17:
Holographic Cloning

In the previous chapter, we explored the possibility of achieving physical immortality via computers. Not only is it clear that the popular notion of "uploading" is a non-starter; but even utilizing the "sub-quantum computer" method when/if it becomes available might not be appealing to many people. Ask yourself: would you *really* want to exist solely inside of a computer, even if that computer was housed within an ambulatory robot? Yes, temporarily inhabiting a robot body would definitely be a fun activity, similar to an amusement park ride; but the phrase "a nice place to visit, but I wouldn't want to *live* there" comes to mind. For those of us who would prefer to retain our *biological* bodies, yet still enjoy an unlimited lifespan, there may be a method that could allow this; and much sooner than the expected time frame for the development of sub-quantum computers. I call this method "holographic cloning."

The roots of holographic cloning began with the work of Dr. Peter Gariaev of the Russian Academy of Sciences. Utilizing a 650nm communications grade HeNe laser, he was able to express the holographic sub-quantum signature of a DNA sample. Since the energy signature persisted long after the 650nm beam was extinguished, and the DNA sample removed from the lab, Gariaev dubbed this effect the "DNA Phantom Effect."

While Gariaev was exploring this newfound effect, a Korean research team headed by Yu V Dzang Kangeng expanded upon Gariaev's work, utilizing a low-power 11-GHz microwave beam rather than a 650nm laser beam. Kangeng's reasoning was that, with a laser beam, you are limited to transparent samples, whereas a microwave beam is capable of penetrating *opaque* samples; meaning that the DNA of *living* creatures could be expressed. Kangeng and his team proceeded to express the holographic sub-quantum signature of a duck and impress it upon a chicken. (See diagram on page 160.) After three days, the chicken began to express duck morphology,[†] at which time the experiment was terminated.

[†] My research assistant refers to the resulting hybrid creature as a "Chuck." It is a running inside joke within the group, so I agreed to grant "The Chuck" a footnote.

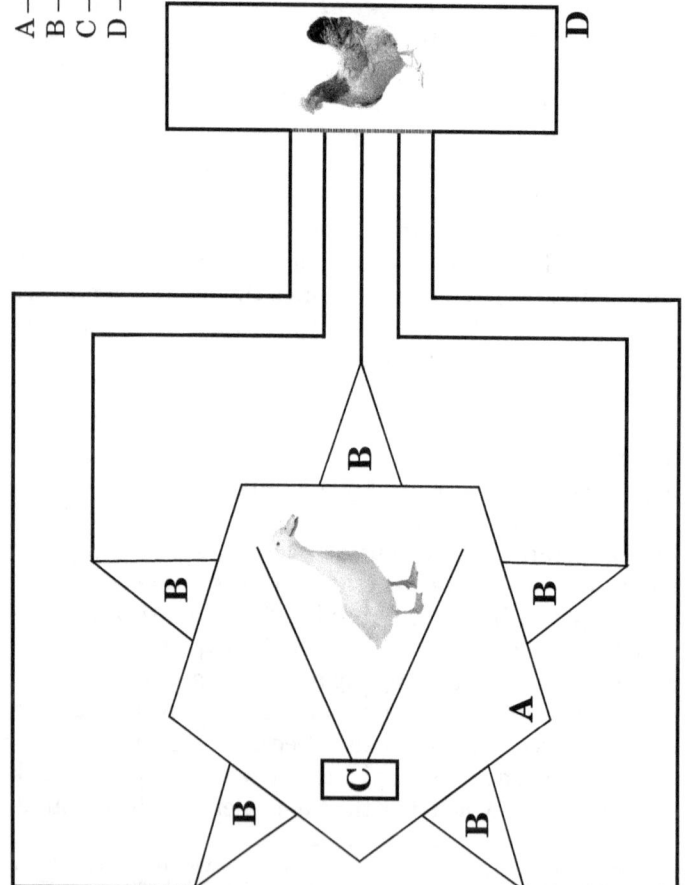

A — Pentagonal Source Enclosure
B — Emitter Cones
C — 11 GHz Beam Generator
D — Target Enclosure

**Kangeng's
Experimental
Configuration**

The chicken was then allowed to reproduce, to determine if the morphological changes persisted. Indeed, the morphological changes were present to varying degrees (as one would expect, considering that the chicken's eggs still had to be fertilized by a second parent) in the majority of the chicken's offspring, meaning that persistent alterations to the chicken's DNA had occurred.

While not as spectacular as his "Chuck" experiment, Kangeng also utilized the same experimental apparatus and configuration to produce plants that were a cross between peanuts and sunflowers.

Not to be outdone (since the original discovery of the "DNA Phantom Effect" was his), Gariaev proceeded to replicate Kangeng's experiment, albeit once again using the 650nm HeNe laser rather than the 11-GHz microwave beam the Korean team had used. This forced Gariaev to use salamander and frog *eggs* (because they are transparent) rather than adult creatures. But whereas Kangeng's team halted their experiment after three days—once duck morphology began to express in the chicken—Gariaev allowed the experiment to continue until the frog eggs had transformed completely into salamander eggs, which then proceeded to develop into perfectly normal adult salamanders.

Pier Luigi Ighina of Italy was able to accomplish a similar feat, utilizing radio waves to transform an apricot tree into an apple tree in only sixteen days. Thus, while the Kangeng experiment only achieved a partial transformation—more of a "splice"—both Gariaev and Ighina accomplished true "holographic cloning," by transforming the target organism into a complete copy of the source organism, likely right down to the holographic sub-quantum signature of the source organism. Unfortunately, since both Gariaev and Ighina utilized life forms incapable of communicating with us, and did not perform comprehensive DNA assays of all test subjects, we cannot confirm the *extent* to which holographic cloning took place. That said, visual inspection *does* reveal that the transformations that occurred as a result of their experiments produced a more extensive transformation than did Kangeng's method (since he halted the experiment after only three days, when duck morphology became visually identifiable in the chicken), so these experiments are definitely worthy of replication. I intend to do precisely that in the not-too-distant future, utilizing a 6-GHz microwave beam (similar to Kangeng's method), but allowing the process to run its course to completion, as did both Gariaev and Ighina. Further, the test subjects will be "Jax mice" (Jax provides genome scanning services for their mice), one of which will be trained in several complicated tasks prior to being utilized as a "source" for holographic cloning. This will not only allow for accurate quantification of the extent to which the target subject's genome was transformed to match that of the Jax-scanned source mouse, but will also provide some measure of proof that the target subject is afterward able to resonate with the holographic sub-quantum mind of said source mouse. Should this procedure prove successful—and the results obtained by both Gariaev and Ighina indicate that it will—one can immediately conceive of how this might be applied to a form of immortality that would allow one to retain their *biological* body, yet still enjoy an unlimited lifespan. Simply have yourself holographically cloned onto a target subject,[†] and enjoy another full life span.

Of course, that begs the question: how can a holographic sub-quantum signature be safely stored? For that, we have a few possibilities. The first comes from the work of the late Dr. Jacques Benveniste. He was able to utilize white noise and identical sensitive

---

[†] I realize that the image which inevitably comes to mind is that of a lunatic "Mad Scientist" abducting people off of the street to be used as "blanks" onto which those wishing to live forever will be holographically cloned. But before torches are lit, pitchforks hefted, and cries of "Madman!" fill the night air; I am advocating nothing of the kind. Once this technology is proven—first with a wide variety of animals, and then upon willing human volunteers; perhaps death row inmates who wish to secure money for their surviving families—it will only be a matter of time before one of Earth's many countries allows the first human clone to be created. (Necessity being, after all, the mother of invention.) It is not much of a leap to envision that "clone farms" would inevitably result, including some chemical methodology that would hasten the development of said clones to adulthood. This has already been conceived, with the idea being to utilize the clones for organ harvesting. In the case of holographic cloning, rather than harvesting *organs* from the clones, the clones would simply be utilized as "blanks" onto which the original "source" human could then be holographically cloned.

microphones to capture a digital recording of the signature of various substances; mainly commercial pharmaceuticals. Benveniste's process created what amounts to a "sound hologram." Much like the technique of a "reference beam" that is necessary to create a *light* hologram, Dr. Benveniste utilized the two microphones to capture the sound that passed through two identical cuvettes: one containing distilled water only, while the other contained distilled water with the substance to be recorded dissolved therein. He would then invert the "reference" sound file (the one that had passed through distilled water only) in MATLAB, and subtract it from the "substance" sound file. What remained was a digital "sound hologram" of the substance. When this sound file was played in close proximity to lab animals, they exhibited the same reaction as to the physical substance, but without the side effects associated with that substance. Needless to say, Dr. Benveniste was not at *all* popular with "Big Pharma." He died of a heart attack not long after word of his research began to be noticed within the public sphere.

Unfortunately, as interesting and suggestive as Dr. Benveniste's technique may be for accomplishing our task—and it does seem to be the first real example of a non-light hologram being recorded—it would be difficult to adapt this method to our needs, for the same reason that Gariaev's original 650nm HeNe laser was inadequate to the task of creating a "Chuck"—laser light requires transparent materials through which it can shine. Similarly, Dr. Benveniste's sonic method may work well for small cuvettes bearing liquid samples and DNA; but remember that, for a complete "encryption key," it is both the subject's DNA *and neural pattern* that must be recorded, to provide the complete holographic sub-quantum signature. And this is where Kangeng's microwave method reigns supreme; if we could but find a way to record the result.

Enter the notion of *microwave holograms*. To date, I am unaware (and believe me, I *have* looked) of any existing medium capable of recording a microwave hologram. That said, it is not difficult to envision a method that could conceivably work. It would likely have been accomplished long ago; but once again, necessity is the mother of invention, and nobody has yet had *need* of a microwave hologram.

Light holograms require film to record them, as film is sensitive to light. If a *microwave*-sensitive, fast-setting (i.e., within a few hours at *most*) polymer could be developed (I have no doubt that either Dow or Dupont could manage this with ease), one could simply replace the target organism in Kangeng's apparatus with a small container of fast-setting, microwave-sensitive polymer. The donor organism would be placed in the "source" chamber as usual, and the microwave beam energized for as long as it takes the polymer to cure. Upon completion of this process, the fully-cured polymer should contain a microwave hologram of the source subject's holographic sub-quantum signature. This could then be used as a "source" within Kangeng's experimental apparatus to create as many holographic clones of that organism as you wish.

This same effect occurs naturally—albeit to a lesser degree—on a daily basis: this is how objects become imprinted (over time) with the holographic sub-quantum signature of a person; enough so that a medium can utilize such an object to "tune-in" to the owner's holographic sub-quantum mind. We are simply creating a process via which this effect can take place over a much shorter period of time, and to a much greater extent.[†]

If a person were to have several microwave holograms of themselves made, and distributed amongst various safe-deposit boxes (or other secure locations)—similar to how one would treat redundant off-site backups of important computer data—this would provide for true physical immortality; and the majority of the technology required to make this happen exists as we speak.

In addition to the holographic cloning aspects of this technology, it may also prove useful in healing the bodies

---

[†] I would be very interested to witness a gifted medium's reaction the first time they are handed one of these microwave holograms.

that we already have. If a microwave hologram is created of your body when it is young, could that recording then be used to irradiate your aging body years later, thereby rejuvenating it? Considering that much of the damage that occurs as we age is due to mutations and transcription errors within our DNA; it just might.

But what if a person is already quite advanced in years by the time this technology is available? Might we, at some point, develop the ability to scan the microwave holograms into a computer (i.e., read them in from the polymer blocks), *edit* them, then re-output them back to a new microwave-sensitive polymer block? I see no reason why not. In fact, some of the methodology to accomplish this has already been developed by Dr. Benveniste. By harvesting a comprehensive set of DNA samples from the subject, and sorting the damaged DNA, one could then create a microwave hologram of *only* the ***damaged*** DNA, scan *that* into the computer, invert it, and subtract it from the previously-scanned, unedited microwave hologram of the aged subject. This would create a "clean" copy of the subject's holographic sub-quantum signature, free from damaged DNA.[†] This could then be re-output to a new microwave-sensitive polymer block, and a new microwave hologram created. Of course, by the time computer technology is that advanced, we will likely be able to eliminate the polymer blocks altogether, as the computer will be able to simply emit the appropriate data directly into the target chamber. It may also be possible by that time to correct—in the computer—any damage that has occurred over time to the subject's neural pattern. (Remember, your neural pattern is also critical to your "encryption key." So if it is damaged, it will impair your ability to fully access your holographic sub-quantum mind.)

And as long as we are already this far down the rabbit hole, extrapolating into the future; let us briefly discuss yet another capability that could potentially be realized via holographic cloning technology: teleportation. I am certain it has already occurred to many of you that,

---

†In effect, this process would act as a sort of "digital rejuvenation."

if "blanks" were on-hand at both ends of what would otherwise be a long journey—say, between Earth and Mars; and eventually, between our solar system and some distant star system—the data for one's holographic sub-quantum signature (i.e., the data that was scanned into the computer from the polymer block bearing the microwave hologram) could simply be transmitted[†] to the destination, and imprinted upon a waiting "blank." Even if it took sixteen days to accomplish this (and the efficiency of the procedure will undoubtedly be improved over time, thereby *reducing* the time required to fully imprint a blank), that certainly beats six months (at current speeds) to Mars, or what could be *hundreds of years* to travel to distant stars.

We can easily see that holographic cloning will be an enabling technology that will not only allow mankind to achieve true physical immortality, but also a wide range of additional capabilities. The equipment required to accomplish holographic cloning already exists, and the basic procedure has been tested. I truly look forward to seeing where this goes.

[†] This would be accomplished via sub-quantum communication systems, naturally; so there would be no delay due to the vast distances involved. Tesla accomplished this almost 100 years ago with his prototype remote control devices, though he referred to it as "scalar communications technology."

Chapter 18:
The Akashic Records

"The day science begins to study non-physical phenomena, it will make more progress in one decade than in all the previous centuries of its existence."
—*Nikola Tesla*

We discussed previously that humanity's *group* holographic sub-quantum mind is analogous to what Carl Jung called the "collective unconscious," and what the Tibetans called the "Akashic Records"—a data depository containing everything that humanity as a whole has experienced and achieved throughout its entire existence.

This data can be accessed in much the same way that mediums access holographic sub-quantum minds other than their own: by bringing *your* mind into resonance with the data you seek. The first and most important aspect of this process is remembering that all information in the universe is contained within *your* holographic sub-quantum mind. You do not need to tap anyone *else's* holographic sub-quantum mind, because *your* mind is embedded within the infinite, universal, sub-quantum hologram; and thus, it contains all information in the universe, in much the same way that even the smallest piece of a film-based hologram is able to reproduce the entire holographic image; though the smaller the pieces are in relation to the entire hologram, the "fuzzier" the resulting image becomes.[†]

This is why most people are not able to consciously access any significant amount of data outside of their own holographic sub-quantum mind. But accessing such information is simpler than most people believe it must be. There is no need to "tune-into" someone *else's* holographic sub-quantum mind: this would only add an unnecessary step, and force you to function over a tenuous connection that relies upon maintaining resonance with another individual's holographic sub-quantum mind. It is much simpler to learn to resolve the "fuzziness" within your *own* holographic sub-quantum mind. And how do you

---

† In physical light-based holography, this is a result of the fixed wavelength of the laser. As the pieces into which a piece of holographic film has been cut become smaller, the ratio of the laser's wavelength to the size of the film decreases, thereby reducing the resolution. It is an inherent physical limitation of the process. Remember that we are not actually talking about a *light*-based hologram here, but a *sub-quantum* holographic data-storage system. As we discussed previously, physical analogies are a guide to understanding, but are not to be taken *literally*. In this case, the physical limitations inherent in a physical system do not apply to the sub-quantum entities being discussed; just as gravity does not exist at the quantum level.

resolve a fuzzy image? *Focus*. This is where a steadfast breathing and meditation routine can pay vast dividends. The stronger and more disciplined your mind, the easier it will be to access the Akashic Records. I am *not* saying that you need to don robes and climb a mountain to meditate for twenty years in a cave. But if you learn a few of the ancient Tibetan breathing and meditation techniques, then apply modern scientific knowledge and insight to the process (i.e., understanding of the science *behind* the mysticism), you will progress at a rate inconceivable to the ancient Tibetans. (More on this in "Chapter 22: Magic(k)")

That said, you will recall our discussion of Raikov and his "artificial reincarnation" method. While Raikov directed his subjects to resonate with the holographic sub-quantum minds of *specific* historical figures— those gifted in either art or music—it would be equally possible to direct the subject to simply access the knowledge they require to perform whatever the task at hand might be, whether painting, or playing a musical instrument. Considering the large pool of talent that is available within the Akashic Records,[†] in addition to the Akashic Records being much easier to access than the specific holographic sub-quantum mind of a deceased person; this would provide a much broader range of talent and data upon which the subject can draw. Combined with modern hypnotic techniques, a simple post-hypnotic suggestion that the subject will retain all abilities and knowledge gained during the hypnotic session would allow for "instant learning" that would persist beyond the bounds of said session.

The subconscious yearning for such "instant learning" capabilities has been at the root of all human achievement in the area of information storage and retrieval. What humans have attempted to accomplish via the creation of various forms of external memory storage technology— first cuneiform tablets and stone blocks; then books; and

---

[†] Since the group holographic sub-quantum mind of humanity contains the sum total of all human achievement to-date; and possibly all future achievements as well. More on this in the "many worlds" and "precognition" chapters.

most recently, digital information storage (including the internet)—is to emulate the storage of data in the infinite holographic sub-quantum infrastructure of the universe: the Akashic Records.

Many indigenous cultures around the world do not have a system of writing; all knowledge is passed-down via "oral tradition." You will note that such cultures have a strong shamanic basis: the people are quite familiar with altered states of consciousness, and accessing data from non-physical sources. Thus, they have felt no need to externalize their data storage, and instead simply pass on the stories of their culture via oral tradition. But in mainstream western culture, lacking the shamanic fundamentals of indigenous cultures, we have relied upon physical data storage methods of varying sorts to preserve and proliferate knowledge.

The only problem with externalizing your knowledge depository, is that tyrannical governments can deny you access to that store of knowledge upon which you have come to depend; and what then? With the oral traditions and shamanic practices utilized by indigenous cultures, nobody can take their knowledge away from them, or deny them access thereto.

From this perspective, the internet is a step-up from books, as it is a decentralized storage medium. For the internet to be taken down, the entire global communications system would have to be compromised. (e.g. By a solar flare, or other such global disaster.) Thus, we can see that this widely-distributed global knowledge bank in digital form is another step towards replicating the Akashic Records in physical form.

This inherent need to externalize and distribute information is simply emulating what mankind as a whole collectively remembers from far back in its history, when the "Akashic Records" were common knowledge; accessible by all. (Think of it as a "cosmic google.") The ancient Hindu scriptures spoke of the "physical world" being only "maya"—illusion. Subconsciously, humanity is trying to reclaim what we instinctively realize is our "real" existence; in this case, by re-creating the Akashic Records in physical form. Perhaps we might be better served by learning to re-connect with the Akashic Records, both technologically (via the aforementioned sub-quantum computers), as well as mentally.

Chapter 19:
The "Many Worlds"
Interpretation Of
Quantum Mechanics

"Everett was before his time. He represents the refusal to relinquish objective explanation. A great deal of harm was done to progress in both physics and philosophy by the abdication of the original purpose of those fields: to explain the world. We got irretrievably bogged down in formalisms, and things were regarded as progress which are not explanatory, and the vacuum was filled by mysticism and religion and every kind of rubbish. Everett is important because he stood out against it."

—*David Deutsch, Oxford University*
(Discussing Hugh Everett, the founder of the
"Many Worlds" interpretation of
Quantum Mechanics

There are two main theories in modern physics to explain how Quantum Mechanics relates to our everyday world. The fundamental question is how a system that has the potential to exist in many different states comes to be in only *one* of those specific states; and what happens to the other *potential* final states once the *actual* final state is determined.

The first theory is known as the "Copenhagen Interpretation," and is most commonly explained via the model generally referred to as "Schrödinger's Cat." Most people are at least passingly familiar with Erwin Schrödinger's thought experiment; but in a nutshell: Schrödinger proposed that, if one were to place a cat in an opaque, sound-proof box along with a canister of poison gas, and an electronic circuit that has a 50/50 chance of triggering the cannister, there would be no way to know if the cat were alive or dead until you opened the box. Per the Copenhagen Interpretation of Quantum Mechanics, until you actually opened the box to check, the cat existed in a "superposed" state wherein it was simultaneously alive *and* dead. (Yes, I know: a ridiculous notion; which is what Einstein and his cohorts were attempting to illustrate when they originally wrote their infamous EPR paper. Alas, EPR was taken seriously, leading to no end of mischief in the years since.) When you finally *do* open the box to determine the cat's fate, whichever possibility did *not* occur is said to be a "collapsed wave function." i.e., You—the observer—by the mere act of observing the experiment, caused the wave function of the unrealized outcome to collapse. If you find yourself quailing at such a bizarre notion, you are not alone. Which is what prompted a physicist named Hugh Everett to formulate a new theory.

In the late 1950s, Everett proposed what came to be known as the "Many Worlds" Interpretation of Quantum Mechanics. This theory held that, rather than the wave functions of possibilities that did not occur *collapsing* upon an experiment's conclusion; instead, multiple parallel universes split off: one for each potential outcome of the experiment. Thus, in Everett's version of

Schrödinger's cat experiment, there would be two parallel universes: one where the cat lived, and one where it died. The larger the number of potential outcomes of an event, the greater the number of parallel universes that would be created.

Now, while this explanation may seem more sensible[†] than the Copenhagen Interpretation/Schrödinger's Cat model, it is not without its flaws. Nature is seldom wasteful, and the notion of an entire universe splitting off over an issue as minor as the life of a cat seems wasteful in the extreme. Further, no mechanism is identified for how this "splitting-off" of parallel universes is meant to occur. We are left to accept a vague, nebulous "somehow." Thus it is that we come to the heart of the matter; and the sub-quantum reality behind it all.

One way to look at the sub-quantum interpretation of the "many worlds" theory is to think of a "Choose Your Own Adventure" interactive book.[††] (Albeit such a book with an infinite number of pages.) When you follow a certain specific path through the book—a path dictated by your choices—the other possible paths through the book do not cease to exist: they are still there, and you *could* go back through the book again, make different choices, and follow one of those alternate paths. Similarly, when a "real world" experiment has a number of possible outcomes (like Schrödinger's Cat; or tossing a pair of dice); when we actually conduct the experiment, and witness only *one* of the possible outcomes, that does not mean that the *other* possible outcomes "cease to exist" (that would be the "collapsing wave function" of the Schrödinger theory); *nor* does it mean that "x" number of separate "parallel universes" were created (as in Everett's "many worlds" interpretation of quantum mechanics). Much like the "Choose Your Own Adventure" book, all of the other possible outcomes of that experiment *do* exist; simply not as "separate universes"—they are

---

[†] Since it provides for the objective reality of the universe, even in the absence of an observer.

[††] If you are unfamiliar with this type of "interactive book," google it; or even go so far as to locate one and read through it. Yes, they are intended for children; but if you never read one as a child, it is an experience that would be worth having. And then pass it on to a child in your life so *they* can have the experience.

all encoded within the infinite holographic sub-quantum infrastructure of *this* universe; the one and *only* universe.

You are likely familiar with the wide variety of available TV-attached video players. (There are a variety of such devices available at your local electronics retailer. Most people have at least heard of Apple's offering in this area, the "Apple TV.") These range from models that can only view internet-based media, to those that have internal hard drives (or to which an external hard drive can be attached), and others that are capable of finding and playing any digital media content on your entire network.

Picture a hypothetical network-attached hard drive of infinite capacity; one that contains all movies, TV shows, documentaries, concerts, etc.—every second of video material ever created in the entire history of television. Now imagine that this hard drive includes all of these shows in fully *interactive* format, allowing you to choose moment-by-moment what happens in the storyline. (You have *unlimited* hard drive space, after all.) This would be the video equivalent of a "Choose Your Own Adventure" book.

Now, extend this analogy to the infinite sub-quantum pixel grid. Here you have a "hard drive" of literally *infinite* capacity, encoded with everything that has ever happened, everything that *could* have happened, in addition to everything that could ever happen in the *future. This* is the true scope of the Akashic Records. In this analogy, your holographic sub-quantum mind is the "TV-attached movie player"—your own personal link to the "network-attached hard drive of infinite capacity." You can see that, rather than the infinite number of "parallel universes" of the traditional "many worlds" interpretation of quantum mechanics; what we really have is an infinite holographic data storage medium—the entire sub-quantum infrastructure of our infinite universe— containing *all possibilities*, which your mind can choose to experience as it sees fit; as if each possibility were a "virtual reality program" that your mind is accessing. And just as you must "suspend disbelief" when you watch a movie (in order to immerse yourself in the movie so you can enjoy it); when you engage with any specific "virtual reality program," you must willingly choose to experience "linear time" from a "physical world" perspective in order to participate in that program. Again, this is simply an analogy; because while we would be limited in the "physical world" to experiencing each "virtual reality" program one-at-a-time; the holographic sub-quantum mind can experience as many as it chooses simultaneously, since linear time as experienced in the "physical world" does not apply outside of this illusory "simulation."

When attempting to grasp the sub-quantum interpretation of the "many worlds" theory, try to picture watching several TVs simultaneously, with each displaying a different movie. While most people would find it a challenge to keep up with several movies at the same time, this is exactly how the holographic sub-quantum mind experiences the various possible branches of one's current life path. It is as if you were capable of simultaneously following every possible path through a "Choose Your Own Adventure" book, or fully interactive movie. We are limited to experiencing each of the possible paths through such a physical book or movie one-at-a-time, due to the limiting nature of the "linear time" we experience here in the "physical world." But your holographic sub-quantum mind is *not* thusly limited, and is fully capable of simultaneously experiencing *all* of the possible variations of your current life; and capable of all of that while *also* remaining conscious across a wide variety of other levels within the infinite spectrum of consciousness.

We already have *light* holograms. We may one day have *microwave* holograms. The Universe is a *mind* hologram. All possible outcomes are already encoded within the infinite sub-quantum infrastructure of our holographic universe; our mind simply acts as the "reference beam" that *expresses* the hologram.

Reflecting back for a moment upon our discussions in the reincarnation chapter, I understand that many people will resist the idea of the ego not surviving death, because they want so badly to believe that the exact person they are right now, will "somehow" survive death. And while the part of them that *does* survive—the holographic sub-quantum mind—will have all of the memories of everything that occurred during their lifetime; it will not be the limited ego consciousness that existed within the "physical world" during said lifetime. That temporary ego construct is like a character one plays in a video game; a fiction. It is utilized for the purpose of having experiences, and is not intended to survive death, as it is an extremely *limited* consciousness.

When you are watching a movie, you know the movie is not real. You realize that these are simply actors playing a part; following a script. Nothing that takes place within the movie is real, from the special effects to the fictitious situations and interplay between the fictional characters. And yet, you willingly suspend disbelief for the purpose of immersing yourself within the story, and enjoying the movie. That state of "movie consciousness" that you willingly—and temporarily—enter allows you to enjoy the movie; but you would hardly wish to remain in that altered state once the movie ends; you would not be able to function in the real world in that state.

Similarly, your state of consciousness while here in the "physical world" is a limited, altered state that you have willingly entered for the purpose of suspending disbelief, allowing you to immerse yourself within *this* "movie." When the "interactive movie" that is your life ends, you return to your normal state: a holographic sub-quantum mind that is simultaneously active across a wider spectrum of consciousness than most people could possibly imagine. That is the "real world," and you could no more function there with your limited "physical world" consciousness than you could function in your everyday life if you remained in "movie consciousness" after leaving the theater.

Upon the conclusion of your current life path, you will still remember all of the thoughts and events from your experience here in the "physical world," just as you remember all of the events from the movies you watch, including your thoughts and emotions about the movie during that immersive experience. But when you are discussing the movie with friends afterward, you are viewing the experience of the movie as something external to yourself, rather than something in which you are currently immersed. Your perspective has changed; and that is how it is supposed to be. You can thus be enriched by the experience you had while watching the movie, and retain forever whatever it is you took away from it; whether that was simply a good laugh, or an adrenaline-fueled ride through an action movie. Those experiences are yours, regardless of the altered perspective from which you view them after the movie has concluded.

Similarly, with your holographic sub-quantum mind being simultaneously conscious across such a wide spectrum of consciousness, this little "blip" of your brief stay in the "physical world" is akin to the two hours you might spend watching a movie. You may enjoy it immensely—even learn much in the process—but its significance should not be exaggerated within the overall scheme of things. The "physical world" is but one experience among a vast range of experiences. This does not diminish the importance of what you experience and learn here; it simply puts this world into perspective. You are fully aware that a movie is simply two hours of illusions on the screen, and that it has no real importance compared with the rest of your life; and yet, you enjoy experiencing the movie. It gives you a break from the "daily grind" in much the same way that a lifetime spent here in the "physical world" is a welcome break for at least a portion of your holographic sub-quantum mind.

So how is it possible for anyone to experience even brief moments of "expanded consciousness" wherein they glimpse the "real world" beyond this world of illusion? I am certain you have had the experience of watching a movie, and having "real world" thoughts intrude; such as "Oh, I have to remember to pick-up eggs on the way home after the movie," or "I need gas on the way home; and

I had better check that low tire." These are momentary intrusions, and usually occur only to the extent that there are "lulls" in the movie, such that your attention is allowed to wander. The more engaging the movie, the fewer such mental interruptions. We would usually refer to that as a "good movie." This is why the "deep thoughts" concerning the nature of our "larger reality" are seldom pondered by those fully immersed within the day-to-day illusion of the "physical world," but rather, by those who have willingly removed themselves from it, specifically in order to explore the greater meaning of existence. They intentionally create their own "lulls" in the "interactive movie" that is their current life. And free will being what it is, they have every right to do so. It is *their* experience to do with as they please.

Since we have addressed the issue of ego, and people retreating from the world in order to contemplate "deep thoughts," this naturally beings to mind the many Eastern religions that engage in this practice. Unfortunately, many Eastern religions also teach that the ego is something to be destroyed in order to obtain "enlightenment." On the contrary: destroying the ego would be like buying the ticket to see a movie, getting your popcorn, drink, etc., and then sitting there, determined *not* to suspend disbelief, resolving instead to remain in "real world mode" throughout the entire movie; constantly reminding yourself that these are simply actors following a script; that the special effects are fake; and that anyone who apparently dies will simply appear in some other movie in a few months. Why would you choose to intentionally destroy the experience for yourself? You bought the ticket, so you should at least try to enjoy the movie. Otherwise, you might as well just leave the building. What would be the point of sitting through a movie you are determined not to enjoy?

Working to "destroy the ego" is a similarly illogical act. It would be like intentionally killing your character in a video game: how are you going to enjoy the game if your character is dead?

In this admittedly illusory "physical world" experience,

the point is to *immerse* yourself in the illusion, and attempt to *enjoy* it. The ego is a necessary component of that process. Now, understanding what the ego truly *is* can actually improve your experience, and make you appreciate it all the more; possibly even allow you to exploit all this world has to offer more fully. But understanding the ego's true nature, and seeking to *destroy* it, are two *very* different things.

Considering the ego, and how it interacts with the "physical world;" let us return to the "video player" analogy discussed earlier in this chapter. Utilizing this video player box, you can choose to watch any of the programs to which it has access; but that does not mean that the other programs that you did *not* choose to watch simply cease to exist. You may choose to watch them at another time, or not at all; but in either case, they will still be there.

When people ponder the "infinite number of parallel universes" concept, they tend to think of it as if the one we are currently experiencing is "real," while all of the *other* possibilities are somehow "less real." But considering that they are *all* equally as illusory, it is not that only *one* of them "actually happened," and the remainder are "collapsed possibilities"—*none* of them "actually happened," as they are *all* illusions. It is simply a matter of which illusion you choose to *experience*.

Since we have already seen that matter and motion are both illusions; by understanding the sub-quantum interpretation of the "many worlds" theory, we can see that the notion of "linear time" is, similarly, an illusion. Thus, in addition to being free to visit any "parallel world" one might wish, we are also free to move forward or backward in what we perceive (while mired within the "physical world") as "linear time." It is similar to "skipping around"—in either direction—through a movie. You can see how this would make time travel possible,[†] without the traditional worry of creating "paradoxes."

[†] Since all of the data for the past, present, or future of the entire infinite universe is encoded within its sub-quantum infrastructure, your holographic sub-quantum mind can choose to experience *any* life path it wishes, in any era of history.

Any changes you make in the "past" of any one specific time line would—if you chose to continue in that time line going forward, in order to experience the result—simply allow you to experience one of the infinite possible alternatives; like watching an alternate version of your favorite movie: one that is *very* similar to the movie you know so well, but also different in various ways.

When I discuss this point at lectures and seminars, I inevitably receive a question regarding the infamous "Grandfather Paradox." So, while the answer is implicit within the explanation above, I will explain this as an example. The "Grandfather Paradox" holds that, were someone to travel back in time, and somehow prevent their grandparents from ever meeting, that would prevent the time traveler from ever having been born; and thus, they would not be able to go back in time and prevent their grandparents from meeting; so they *would* be born, and go back in time—you can see where the causal loop beings here; the "paradox." But within the context of *either* interpretation of the "many worlds" theory, were the person to travel back in time and prevent their grandparents from meeting, all that would occur is that the time traveler would then experience a "parallel world" in which they were never born; but that does not mean that their *original* branch—the one from whence they came—would suddenly cease to exist: they have simply taken an action that will cause them to experience a *different* branch than the one into which they were born. No paradox here; move along.

In previous chapters, we discussed how Dr. Raikov's procedure, as well as the Akashic Records, could be utilized for "instant learning;" at least, for accessing knowledge and abilities that anyone in the history of humanity had already learned or accomplished. For painting or playing musical instruments, this would prove quite useful, as well as for bringing students up-to-speed on the vast body of knowledge that the human race as a whole has already accumulated. But what of "instant learning" of knowledge that has yet to be discovered? The sub-quantum interpretation of the "many worlds" theory holds that potential, since we can access all possibilities.

For instance, assume you are an architect, have always been an architect, and since you were very young, you had always wanted to be an architect. The majority of your other potential life paths would consist of you in careers, with varying degrees of success, as an architect. Certainly, many would contain "failed" scenarios where you end-up washing dishes for a living; but there would likely not be any where you were a brain surgeon. Thus, were you able to consciously access your holographic sub-quantum mind, and all of the knowledge you possesses across all of your potential life paths[†], there would be quite a bit of data on architecture that could help you in your "here and now" career: pitfalls to avoid, inspired ideas that have not yet occurred to you; but nothing in the way of macro- and micro-neuroanatomy as would be required of a brain surgeon.

But were you to decide that brain surgery was something in which you might be interested, and you resolved to learn everything you could about the subject (possibly including entering medical school); your holographic sub-quantum mind would immediately be populated will all of the data from every potential life path in which you succeed to varying degrees (i.e., actually become a brain surgeon), as well as the inevitable failures. Thus, you would be able to acquire the knowledge and skill set required to become a brain surgeon, as well as data regarding what your life could be like along those life paths, without ever having enrolled for a single class. Thus, with your *own* holographic sub-quantum mind *alone*, you already have unlimited knowledge at your disposal.

Whereas with the "Choose Your Own Adventure"

[†] Since it is easier to access your *own* holographic sub-quantum mind than that of another person; and the effects are longer-lasting. People who underwent Dr. Raikov's procedure found that, following the hypnotic periods wherein they accessed the holographic sub-quantum minds of famous musicians, artists, etc., they noticed *improvement* of their skills in those areas; but nothing akin to the talent they exhibited while under hypnosis. Similarly, people adept at accessing the Akashic Records—such as Edgar Cayce—were unable to retain what they had accessed once they exited the altered state required to perform their feats. When accessing your *own* holographic sub-quantum mind's full store of data—the data available across all of your potential life paths—the knowledge and abilities gained would be much longer-lasting; possibly permanent, at least with repetition and reinforcement.

book, you can only follow one branch at a time; in the "real world" of the holographic sub-quantum mind, you do not even have to "choose one"—your holographic sub-quantum mind can experience *multiple* branches, and be active within them all *simultaneously*. Just as your holographic sub-quantum mind is simultaneously active across many levels within the infinite spectrum of consciousness; it can also be simultaneously active across many "parallel worlds" *within* a small segment of that infinite spectrum; in this example, the portion that we call the "physical world." (Nature is efficient that way.) Just as your mind is distributed across the infinite holographic sub-quantum infrastructure of the universe; so, too, is its attention distributed across a vast number of "parallel worlds" within the "physical world." Thus, the holographic sub-quantum mind is the ultimate in multi-tasking.[†]

Since everything we see in the so-called "physical world" is an illusion, it is analogous to the video game elements we discussed in chapter three, interacting with each other according to pre-defined rules established within the programming code of the video game. In our illusory physical world, objects must obey the laws of physics, where the concept of "cause and effect" reigns supreme. But when you comprehend the larger universe, you see that cause and effect does not apply to how one chooses to experience various "physical world" scenarios from the perspective of the holographic sub-quantum mind, any more than the rules of any given video game can control your behavior in the "real world." And just as the rules within any given video game apply *only* to that particular video game; so, too, do the laws of physics and cause and effect apply only to the "physical world" while we are *here*. For the duration of the time that you are experiencing any given situation *within* the physical world, you must "obey the rules," exactly as you must in a video game. (Unless you enable "cheat codes." More

---

[†] We discussed in previous chapters how errors in communication between the physical brain-body system and one's holographic sub-quantum mind are often classified as "mental illness." We can immediately see how people who have difficulty mentally segregating the multiple "parallel worlds" across which their holographic sub-quantum mind is simultaneously active could find themselves being classified insane, and locked away in an asylum; or at the very least, heavily medicated.

on that in "Chapter 22: Magic(k)." But once you have experienced a given scenario within the "physical world," you can then choose to experience any *other* scenario—related or not—in any *order* that you wish; much like "skipping around" in a movie. If you choose to watch the movie in the normal way (as you would if you were in a movie theater), you will experience the illusion of linear time within the movie's story. But you are also free to skip around as much as you wish; say, for instance, to watch your favorite scenes in a move that you have watched many times.

In much the same way, you can experience all aspects of any given physical world scenario by "skipping around" to the various "parallel world" variations of that series of events. The laws of cause and effect do not follow you outside of the various scenarios within the "physical world," and nothing that you do in *any* of the various scenarios you may inhabit can negatively affect your holographic sub-quantum mind. Your involvement in any variety of "physical world" scenarios can only add experiences to the store of knowledge within your holographic sub-quantum mind. And to a mind that has access to all information in the universe via the Akashic Records, *experience* is everything; the difference between reading about an activity, and actually *taking part* in that activity.

Reflect back upon various episodes from your life. Some of the events and situations you experienced may have felt downright tragic at the time; but you can often laugh when you look back on such episodes, since your perspective has changed. Since the holographic sub-quantum mind can simultaneously experience a multitude of differing scenarios surrounding events in your life, that perspective is widened even further, as *all* possible outcomes have been explored, for *every single event* of your life.

So when you find yourself in a situation in this life that is not to your liking, and you wonder why your holographic sub-quantum mind would torture you so; realize that, from the perspective of your "real self"—the holographic sub-quantum mind—not only are *all* of the possible variations of your life merely a "minor diversion" among the many levels of the infinite spectrum of consciousness across which it is simultaneously active; but each of the life paths it chooses to explore is simply one potential path among many; like going back through a "Choose Your Own Adventure" book numerous times, and exploring all of the paths you did not take the *first* time through the book. But whereas you had to back-up and go through those multiple alternate paths one-by-one, your holographic sub-quantum mind can experience all of your possible life paths *simultaneously*; and your current experiences are but one possible path among many. So your holographic sub-quantum mind is not choosing to "torture" you as its hapless "video game character;" it is simply exploring all possible paths your life could have taken.

Much as later in life, when you tell your life's stories to new friends (e.g., about the time something happened that may have seemed horrible at the time; but now, you can look back on it and either laugh, or at least reflect upon what you *learned* from the experience), your holographic sub-quantum mind can look back upon *all* of the many versions of your life, and reflect upon how the various life paths differed based upon what choices you made, and the resulting events you experienced as a result. These events take on a different meaning from the perspective of your holographic sub-quantum mind, just as past events do to your current ego consciousness here in the illusory "physical world."

In addition to experiencing multiple variations of a life simultaneously, your holographic sub-quantum mind might choose to go back and re-experience some of your favorite moments from various lifetimes (remember, in the *real* world of the infinite sub-quantum infrastructure of the universe, there is no "linear time" as we experience it here in the "physical world"), similar to "skipping around" in a DVD or Blu-Ray that you have seen before, so you can watch your favorite scenes. But again, as with the "Choose Your Own Adventure" book analogy, whereas we in the "physical world" are forced to experience the various scenarios one-by-one; your holographic sub-quantum mind can experience any number of them simultaneously.

Many people, upon considering the sub-quantum interpretation of the "many worlds" theory, perceive this to mean that there exists, in reality, no such thing as "free will;" i.e., that everything is "pre-destined," thereby eliminating free will. But nothing could be further from the truth.

In fact, the sub-quantum interpretation of the "many worlds" theory provides for the greatest possible degree of free will. Think back to the "video player" analogy. If you attach a hard drive to such a video player, with said hard drive being capable of holding, say, 1,000 movies; you then have the "free will" to choose which of those 1,000 movies you would like to watch at any given time. A plethora of choices, to be sure; but still a *finite* (and thus, *limited*) number of choices. Now, imagine connecting that video player to a network-attached storage device containing 20,000 movies, in addition to a variety of TV shows, documentaries, concerts, etc.—a *much* larger selection of choices than with the attached hard drive. The number of available choices is still finite, since even the network-attached storage device's hard drive space is finite; but it provides significantly greater variety than does the physically attached single hard drive capable of containing a mere 1,000 movies. Thus, you have vastly more "free will" than you did with the single hard drive.

Now imagine that you could connect to a hypothetical "cloud server" on the internet that contained every piece of video footage of any sort that had ever been produced in the entire history of television, in addition to every piece of video footage that *could* ever be made; across every intelligent species in the universe who might ever invent a visual display entertainment device similar enough in structure to your television. In an infinite universe, this would give you a truly *infinite* choice of viewing material.

A television screen is inherently capable of displaying every video of any type ever made. It has the pixel structure to display anything capable of being displayed on a device of its type. This provides for a near infinite variety of visual imagery; and it is all inherent within the limited pixel grid of a television screen. Before a movie is even *produced*, that television screen has the potential to display it; but that does not reduce or remove the "free will" of the movie producer. Similarly, the sub-quantum pixel grid already has the potential to display any possible version of the "physical world;" but that in no way reduces or eliminates the free will of anyone who chooses to participate in any of those potential worlds: they all have complete free will (within the confines of the laws of physics) while they reside within any of those "physical worlds." That is the only restriction made upon free will: you have to "play by the rules," just as when you play a video game. (Again, unless you enable "cheats." See "Chapter 22: Magic(k).")

Similarly, any book that has ever been or will ever be written is inherent within the pages of a dictionary. Does that reduce or remove the free will of any author who ever chooses to write a book? One's choices and free will are not limited simply because the medium via which one expresses creativity is of limited extent. This applies equally to writing a book, making a movie; or any actions we choose to take in any of the possible "parallel worlds."

In an infinite universe, containing an infinite number of what are referred to (within the standard "many worlds" interpretation of quantum mechanics) as "parallel universes," your holographic sub-quantum mind has—quite *literally*—an *infinite* variety of experiences from which to choose; and that is only within what we call the "physical world." Add to that the many other levels of the infinite spectrum of consciousness across which one's holographic sub-quantum mind is active, and you begin to see how, rather than *limiting* our free will, having every possibility encoded within the infinite holographic sub-quantum infrastructure of the universe (the sub-quantum "pixel grid") actually provides us with an *infinite* amount of free will.

And yet, people persist in asking "But doesn't this all mean that everything is pre-destined; that there is really no such thing as free will?" That is no more true than it would be to say that a book an author just wrote was "pre-destined," simply because all of the words he used to write it were already in the dictionary.

In any book an author might write, they are utilizing their own free will and imagination, but must make use of the "building blocks" of words; which are collected in a dictionary. The fact that every word in an author's book is already contained within the pages of a dictionary in no way limits their free will or imagination: they are free to combine those words in any way they choose.

We discussed in chapter three how the sub-quantum wave packets are like an alphabet, producing the "words" that comprise the "fundamental particles" of the Standard Model of particle physics. Thus, the particle zoo of the Standard Model is essentially a dictionary containing all of the "building blocks" of matter.

If you think of the infinite possibilities encoded within the sub-quantum infrastructure of the universe as a "dictionary of *events*,"† you can see how this analogy applies to a person's free will, and their current path through the "physical world." Yes, all possible events, actions, and even thoughts may already be encoded within this sub-quantum "dictionary;" but every mind is free to make whatever choices it wishes from those infinite possibilities, and therefore experience an infinite variety of life paths. This is the very definition of free will: the freedom to make your own choices, and experience the consequences of those choices.

By this point, it should be quite clear that, simply because any choice one could possibly make has been accounted for, does not mean that your life is

---

† Think of this "dictionary of events" idea in relation to a video game: any action you can possibly perform within that game is already pre-programmed into the code for the game. And yet, you still have fun playing the game. Similarly, once computers are powerful enough to make true "virtual reality" a practical reality, people who enter such virtual worlds will have much greater ability to interact with their environment than exists with current video games; and yet, any action they could possibly take within that world will still have to be pre-programmed into the code for that virtual world, else they could not take said action. Now, remember that the video game/virtual reality examples are simply *analogies*. In the "real world," rather than our actions being limited by a finite piece of computer programming that is capable of accounting for a finite number of actions, *all* possibilities are already encoded within the *infinite* sub-quantum infrastructure of the universe. This provides us with a level of freedom impossible to achieve in any video game or virtual reality construct.

"pre-destined," or that you have been deprived of "free will." In the chess game we call life, the fact that all possible moves are already known does not mean we cannot have fun playing the game.

Approaching this from a slightly different angle for a moment, let us consider the primary colors—the additive and the subtractive primaries, depending upon whether you are working with television screens or physical pigments. If you give an artist a blank canvas, with a brush and a limited number of pigments, they can create every color that exists within the human visual spectrum. And yet, you can provide the exact same materials to a million different artists, and each of them will create a different image. They were all provided with the same limited number of pigments, but they each created something unique based upon their individual talents and choices. It seems that providing the artists with a paltry handful of pigments (a *very* limited pool of choices) did not curtail their creativity—their "free will"—in the slightest.

Similarly, a television screen contains pixels of only three colors: red, green, and blue. And yet, those three colors can combine to make all of the video imagery that has ever been produced in the history of television; and all that ever *could* be made. Advances in video display technology continues to improve the *resolution* of television screens, but the human eye only has red, green, and blue receptors; so the red, green, and blue pixels in a television screen produce the entire visible spectrum, regardless of the resolution of the images displayed. Those red, green, and blue pixels can produce roughly 16.7 million possible colors. That is 16.7 million colors from only *three* colors, simply by combining those three "primary colors" in unique ways.

Now picture the infinite sub-quantum infrastructure of the universe. Inherent within that infrastructure is a truly infinite variety of possibilities. Much as the red, green, and blue pixels in a television screen can create all of the colors the human eye can see; similarly, the infinite number of possibilities inherent within the sub-quantum infrastructure of the universe is available, from which you can create whichever reality you choose.

Your holographic sub-quantum mind is encoded within the same holographic infrastructure as the rest of the infinite universe. So whereas with a television, the human eye and the television screen are (within the context of the "physical world") *separate*; our minds and the sub-quantum infrastructure of the universe are all part of the same infinite universal hologram. Within the infrastructure of the infinite sub-quantum pixel grid, everything is encoded: everything that has ever been, or could ever be.

Once you say that everything is "already encoded," people tend to interpret that as meaning that everything has "already been done." But just as with the analogy of a television screen, it is not that everything that *could* be made *has* been made; it is simply that anything that *could* be made, is already inherent within the pixel grid of that television screen.

From our limited perspective here in the "physical world," thinking in linear time, we say that an event "has not happened yet." But within the infinite sub-quantum infrastructure of the universe, every possibility is accounted for; everything that ever *has* been, or ever *could* be, is encoded. Just as any piano melody you could ever compose is already implicit within the keys of a piano. When you write a new melody, you are *expressing* one of the many possibilities that are implicit within those keys. In much the same way, any event that could ever occur, any experience that could ever be had, and any thought that could ever possibly be conceived, is already implicit within the infinite sub-quantum infrastructure of the universe. This does not diminish the person who has an original thought, invents something new, or even composes a new melody. The fact that these things were implicit does not mean that their expression is inevitable: the possibility can exist without ever being realized. Thus, not only is *free will* preserved, but the value of creative thought, artistic expression, and scientific inventiveness as well.

Consider the world's most powerful supercomputer

playing chess against a human chess master. Before its human opponent ever makes a single move, the computer can calculate every move that could possibly be made. Once the human opponent actually *makes* the first move, the computer can then calculate all of the remaining possible moves from that point forward. And so on, until the game concludes. Much like the aforementioned "Choose Your Own Adventure" book: only *one* path is taken, though a multitude of other paths are also encoded within the book; paths that you *could* have been taken. Similarly, the computer already has encoded within its programming every single move and combination of moves that could ever possibly be made. It can run simulations demonstrating each and every possible combination of moves that could play out. So, even though only one *actual* scenario plays out in each individual game, the other possibilities are still encoded within the computer. The variable here: the computer does not know which move the human opponent will make next.

Within the infinite sub-quantum infrastructure of the universe, all possibilities are encoded, just as all possible chess moves are encoded within the chess-playing computer. But unlike the computer, which is limited to playing a single game against a single opponent, and therefore only experiencing *one* of the many possible paths through the chess game; our holographic sub-quantum minds are *vastly* more advanced than a computer program, and are capable of simultaneously experiencing *many* different paths—akin to being able to follow several paths through the "Choose Your Own Adventure" book simultaneously. So while all possibilities are already encoded, we are not obliged to follow any *specific* path. Thus, free will is preserved.

I realize it may seem that I am belaboring the point regarding the "pre-destiny/free will" issue; but during the Q&A sessions following my lectures and seminars, this topic seems to generate an inordinate number of questions; which indicates that the querents do not fully grasp the sub-quantum interpretation of the "many worlds" theory. But whereas at a lecture or seminar, attendees are able to ask questions, allowing me the opportunity to explain further (often at length), until every last person present comprehends this admittedly difficult concept; this type of interactive exchange is not possible in written form. Thus, I have attempted to thoroughly address the issue here, as proper understanding of this topic is crucial to apprehending the subjects that will be addressed in subsequent chapters.

In addition to the issue of "free will," another commonly expressed concern regarding the sub-quantum interpretation of the "many worlds" theory is the fear that it could lead to an epidemic of apathy; that people would believe that nothing they could ever do would matter anyway, so "why bother?"

Does a movie that you watch for fun really "matter?" In the larger scheme of your life, will watching that movie really change anything? And yet, you watch the movie anyway. Why? For enjoyment; for a bit of escapism. It might not "matter" in any larger sense; but you enjoy the *experience*, and that is all that truly matters.

If you do not begrudge yourself watching a *movie* for a bit of relaxation and escapist fun now and then, why would you begrudge yourself a brief vacation in the "physical world?"

But what of people who perceive themselves to be stuck living "bad lives?" Much like a "bad movie" that fails to engage you—where you find yourself thinking about "real world" issues during said movie—this is why the "deep thoughts" about the nature of reality are generally contemplated by people who perceive themselves as living "less than ideal" lives. And just as with a movie, you never really know beforehand if it will be a *good* movie, or a terrible movie; with life, you never know how the possibilities will branch. So you sit back, and take the ride.

But whether a person's current experience in the "physical world" is perceived by them as being a "good life" or a "bad life," it is much like with a movie:

even if it was terrible, you will still have fun dissecting it with your friends later, if only to discuss how awful it was. Either way, it was an *experience*; and that is what the "physical world" is all about.

If people truly comprehend this, there is no reason for them to become apathetic. Whatever the perceived quality of their current life—how good or bad the "movie" may be—they can rest assured that they will have a laugh over it later; or at the very least, have some interesting events and situations to discuss with their friends. (There is a *second* reason that apathy would be an unlikely result of exploring the sub-quantum interpretation of the "many worlds" theory; but since it is the *sole* moderating factor in the next topic, we will deal with it below.)

On the flip-side, there may be those who believe that understanding the sub-quantum interpretation of the "many worlds" theory has the potential to "end all war," because people would see how futile such petty squabbles are. Unfortunately, this would be as unlikely as apathy, and for a similar reason: people are creatures of habit.[†] While a certain small percentage of the Earth's inhabitants (even if a million people were to read this book and grasp its contents, that is an infinitesimal percentage of the total population of this planet) may eventually come to understand the sub-quantum interpretation of the "many worlds" theory; and while it *may* help them to take life a bit less seriously; in the end, old habits die hard. Mired as they are within whatever life path they were following *before* coming to understand the sub-quantum interpretation of the "many worlds" theory, people will easily fall back into old habits nearly immediately. Yes, in the back of their minds, this new understanding may temper some of their negative reactions to some degree; but even were the entire population of Earth to eventually comprehend these realizations; in the end, it would make little difference.

Never underestimate the power of mental inertia. In much the same way that people can go to a movie

---

[†] How many of you have purchased a new piece of exercise equipment, only to use it for a few weeks, then have it gather dust in a corner, or become an impromptu storage space for miscellaneous debris?

theater, and temporarily suspend disbelief in order to *enjoy* said movie; once they *leave* the theater, they quickly revert to their "real world" personas, and return to life's daily needs and problems. Granted, some movies produce a more profound and lasting effect than others; but even these rare gems quickly fade into the background when bills must be paid, and all of life's other concerns pull one inexorably back into the mire. And there is nothing wrong with that: it is, after all, what we came here for.

But perhaps having the truth tucked somewhere in the back of our minds (and possibly experiencing, as a result, a few recurring "moments of clarity" here and there), while it will not make war and religion obsolete, may at least have the potential to moderate our more negative and violent reactions to life's daily insults. It may not be "world peace," but every little bit certainly helps. (Especially in light of the "Maharishi Effect," which we will discuss in "Chapter 22: Magic(k).")

In the end, the indefatigable pragmatism (and mental inertia) of the human condition will tend to average-out any psychological effects an understanding of the sub-quantum interpretation of the "many worlds" theory might otherwise produce. While this may be lamentable from a scientific standpoint (for it is one of the reasons that change in science takes place so slowly, with new ideas sometimes requiring generations to take root), it will at least ameliorate any of the potential negative psychological effects discussed above.

Chapter 20:
Precognition

"Life can only be understood backwards; but it must be lived forwards."
                              —*Søren Kierkegaard*

The material in the previous chapter will undoubtedly raise the question of why it is so difficult to look *forward* with any level of accuracy along one's current life path. This is because precognition is completely a matter of *probability*. In any attempt to discern the future, one obviously wishes only to reveal the future of their *current path*. The fewer the total number of possible outcomes, the more likely you will be able to obtain an accurate picture of the future. But the higher the number of possible outcomes, and the greater the extent to which events can be influenced by minor variables, the more difficult it will be to determine what the *actual* future of your current life path will be. It is all a matter of probabilities.

The phrase "hindsight is 20/20" definitely applies here. Upon looking *backward* along your current life path, you can easily review all of the twists-and-turns that brought you to your present position; the path you chose through the "Choose Your Own Adventure" book *this* time through. When you look back upon the path you followed to arrive at your present position, no matter how convoluted that path may be, it *is* the path you followed; at least, in this particular "parallel world." When you look *backward* along a given path, you will see the events and choices that comprise *that path*.

It is when you attempt to look *forward* that things become more difficult, as you really have no way of knowing for certain how events will play out; which of the many possible branches your life path will actually follow. And that is the crux of the matter: certainty.

Some things are fairly predictable—the Sun and Moon rising and setting, for instance—while others are less so. Take, for example, playing the lottery. Imagine trying to account for all of the variables involved with a bin full of numbered ping-pong balls being buffeted about; there is truly *no way* to predict which of those tiny numbered balls will end-up in the precise position to be drawn up through the selection tube. The greater the number of variables involved in a given event's outcome, the less predictable said outcome will be. Thus, while one can always explore the many *possible* futures that *could* branch-off from one's current position along their life path, there is no way to be certain *which* of those many paths you will find yourself following.

This is why precognition is so unreliable. While any good scientist can predict future events within certain bounds—such as the above-mentioned rising and setting times of the Sun and Moon—precognition generally involves itself with future events that cannot be sufficiently quantified via scientific means to produce solid predictions. Thus, those who are capable of accessing the Akashic Records in order to peer forward will often make predictions based upon the most predominant events witnessed across the many possible future branches; i.e., events that are seen in *many* of the possible future branches. Precognition may never allow one to predict the Lotto numbers, due to the tremendous variability (quite intentional, that) inherent in the process; but something such as a major earthquake— an event that will occur across a large number of the possible future branches—is a safer bet. If a "seer" (one who has developed their precognitive abilities) discerns that such an event is imminent (usually long before seismologists receive any warning via their instruments), they will convey this prediction to any who will listen; and quite often, they are spot-on.

Naturally, nay-sayers and self-proclaimed "skeptics" (most of whom, as we discussed previously, are actually nothing of the sort: merely debunkers) will state that such predictions are nothing more than "luck;" that since such predictions are made so frequently, they are "bound to be right once in a while." i.e., They chalk it up to "coincidence." Sadly, the typical argument of such a debunker is: "If someone could really tell the future, why wouldn't they use that ability to win the lottery?" The debunkers obviously fail to grasp the concept of probability as discussed above. A decent statistics professor would do them a world of good.

Since most of us are used to experiencing time as a one-way linear flow, this is how our minds become conditioned to perceive reality. That is why most "premonitions" occur in *dreams*, where we are more open to a non-linear flow of events. Thus, for those who seek to develop their innate precognitive abilities (for *all* of us are inherently capable of accessing the Akashic Records), the best way to proceed is by first perfecting the ability to remember one's dreams. That accomplished, learning to have *lucid* dreams would be the next step, thereby allowing you to *steer* your dreams towards specific future events that are of interest to you. Thereafter, self-hypnosis and meditation may eventually allow you to access your precognitive faculties while you are awake. It depends solely upon how much effort you are willing to put into the process.

Developing your innate precognitive abilities may never allow you to win the Lotto; but by utilizing the probability analysis method as discussed above (i.e., whichever events you see occurring across many of the possible future branches), such abilities could assist you in navigating the many potential branches along your current life path.

Chapter 21:
Astral Projection &
Remote Viewing

*"The greatest illusion
is that mankind has limitations."*
—Robert A. Monroe

In this day and age, nearly everyone has heard of the "near death experience" (NDE). While such a state as NDE is triggered by a person having temporarily died, astral projection occurs every night while we sleep. As you are falling asleep, you go through what is known as the "hypnagogic state." It is during this state that your mind processes the events of the day. You pass through a *second* hypnagogic state as you transition upward from sleep into wakefulness. In-between these two hypnagogic states, you alternate between various levels of REM and non-REM sleep. REM sleep is the state wherein you dream; and the deepest of the non-REM states is when astral projection occurs. (Astral projection occurs during the portion of sleep wherein you are closest to death, *not* during REM episodes as is commonly believed.) This is when your "physical world" ego consciousness is least active.

During the hypnagogic state prior to waking, your mind processes what you experienced while you were asleep: a combination of dreaming, and impressions from the various other levels across which your holographic sub-quantum mind is active. Such impressions are commonly referred to as "astral projections." Thus, it is not that one must *learn* to astral project, but rather, to remain *conscious* during an astral projection.

"Remote Viewing" is simply a more limited form of astral projection that was developed mainly for use by military/intelligence agencies. With remote viewing, rather than ranging across other levels within the infinite spectrum of consciousness, the remote viewer remains within the "physical world," focusing their attention upon a remote location for the purpose of gathering data about that location.

In all three of the above cases—NDE, astral projection, and remote viewing—the same basic phenomenon is at work, simply under varying circumstances. In essence, your "physical world" ego consciousness momentarily loosens its rigid control over how perception is "supposed" (from its perspective) to occur, and your focus is temporarily able to roam a bit more freely. Consider, for example, the type of astral projection that occurs every night while you sleep. It is not that your mind actually "leaves the body," for as we have discussed previously, your mind was never really "in the body" to *begin* with. Rather, what occurs is that, while your "physical body" is asleep, your holographic sub-quantum mind remains conscious across many other levels within the infinite spectrum of consciousness. Often, upon waking, you recall portions of what your holographic sub-quantum mind perceived on a portion of those levels: the levels that are similar enough to the "physical world" for your limited ego consciousness to comprehend; i.e., those for which your "day-to-day" self has a frame of reference. Thus we can see that the very term "astral projection" is a misnomer, as you never really "project" *anywhere.* It is all a matter of perception.

This explains why people find it so difficult to learn to *consciously* astral project: by thinking of "leaving their body," they are approaching the practice from the completely wrong angle. There have been *many* books written about astral projection—Robert Monroe's being among the best—so this chapter is not intended as a primer on astral projection; nor to prove that the phenomenon is "real;" nor as a "how to" guide. The goal is to understand the phenomenon of astral projection in relation to the holographic sub-quantum mind.

For one's holographic sub-quantum mind to be active *only* within the "physical world" would be as wasteful as having access to the hypothetical "infinite hard drive" discussed previously—the one containing every second of video material ever produced in the history of television—and only being permitted to watch movies of a single genre, and only those produced during a handful of years. Nature is never that wasteful; which is why our holographic sub-quantum minds are capable of being simultaneously active across multiple levels of the infinite spectrum of consciousness.

Similarly, astral projection is not limited to a single "Astral Plane"—you are able to access multiple levels while astral projecting; it is simply that we are only able to reconcile relatively few of those levels within the parameters of our "physical world" ego consciousness. Regular practice can extend one's range; but there are inherent limits to how much our ego consciousness can comprehend. Beyond a certain point, the experiences would be so far removed from "physical world" experiences, that there would be no frame of reference that would allow your ego consciousness to relate.

For this reason, the astral levels that people discuss most frequently are those bearing the greatest similarity to the "physical world," as the experiences there are easier to relate to one's day-to-day existence.

Thus, it is not that the so-called "astral planes" are *all* that exists beyond the "physical world;" it is simply that they are the most familiar to us. All of the various "astral planes" are still completely contained within the infinite sub-quantum pixel grid. We tend to view these levels as "separate," as they seem to be thus from our "physical world" perspective. (There is that "matter-centric" thinking again.) But in reality, they are (much like our "physical world") nothing more than illusory manifestations of the sub-quantum pixel grid; simply additional small fragments of the infinite spectrum of consciousness.

Many of the Eastern religions and schools of mysticism focus heavily upon astral projection as a tool of "enlightenment." With each "level" an aspirant "ascends," they think "Oooh, I'm really getting somewhere now!" But this is nothing more than an illusion. No level is of any more intrinsic value than the next.

The *true* "enlightenment" at the "peak" of their development (which will arrive at different levels for each individual) will be when they finally see the "big picture," and realize that it is all an illusion: that each level is as good as the next, depending solely upon what it is you wish to experience at any given moment. No room for "snobs" in enlightenment.

It is simply a matter of *focus*. While in the "physical world," one's focus is *very* narrow: a "this is all there is" limiting thought structure. (This is the "suspending disbelief in order to participate" aspect that we discussed in previous chapters.)

But as the above-mentioned aspirants continue to experience the "astral realms," their focus may eventually widen to understand that there are an *infinite* number of levels within the overall spectrum of consciousness; and yet, many such aspirants maintain the false belief that the "higher" they ascend, the closer to "enlightenment" they will be. But this is nothing more than a stick with a carrot dangling from its end: the "reward" is illusory. The only "enlightenment" to be had is the realization that, fundamentally, all of the levels are the *same*. Yes, they may have different attributes that appeal to various people in a number of ways; but one level is equally as illusory as the next.

It is similar to a TV with many stations available: one group of people may agree to watch one type of show, while the next group chooses something entirely different. But the TV itself did not change; and while the programs the two different groups chose may have been *very* different with regard to content, they were also—at the most fundamental level—nothing more than pixels turning on-and-off in various patterns: an illusion. And thus, essentially the same.

One's holographic sub-quantum mind is *always* distributed across the entire infinite sub-quantum infrastructure of the universe: it is simply a matter of where you choose to *focus* at any given moment. It is similar to "un-stirring the glycerine" (to borrow David Bohm's example) in order to reveal the drop of ink that was previously "enfolded" within the glycerine by the initial stirring. When you *focus* in a given area (whether within a single world, as with remote viewing and NDEs; or across multiple levels, as with astral projection), that is where your consciousness *appears* to be located. But that is the illusion: you were always non-locally distributed, no matter *where* you were previously focusing. This is the "enlightenment" that aspirants seek; something that every holographic sub-quantum mind already *knows*. But to achieve this knowledge while still within the "physical world," and thus, possessing only limited ego consciousness; that is the trick.

Chapter 22:
Magic(k)

*"Any sufficiently advanced technology is
indistinguishable from magic."*
                    — Arthur C. Clarke

First off, I want to stress that this chapter is *not* about "proving" that "magic(k)" exists, as innumerable volumes have already been written for that purpose.[†] Our objective here is to comprehend *how* such effects are possible.

In previous chapters, we have discussed several topics (telepathy, precognition, mediumship, psychometry, etc.) that most people might consider to be "magic(k);" but by the strictest definition, such abilities are simply different ways in which the mind can *perceive* the world. What is meant by "magic(k)" is producing tangible effects within the "physical world" in the absence of any standard "physical world" causative agents;[††] i.e., affecting the world with the mind alone. That is why telepathy, precognition, etc. were granted their *own* chapters: they are merely methods of transferring *information*, with no tangible effects upon the "physical world"—the mind *perceiving* rather than *influencing*. That is precisely why "psionics" is covered within *this* chapter, rather than receiving a chapter of its own: while some psionic devices *can* be used simply for information gathering (diving rods, Ouija boards, pendulums, etc.), other psionic devices can be used to aid the mind in directly *affecting* the "physical world."

For the remainder of this chapter, we will discuss exactly how "magic(k)" can occur. Just as those who are advanced enough with computers to enable "root" capabilities in a UNIX terminal can perform feats that "mere mortals" (i.e., "end users" in computer industry parlance) cannot comprehend; "magic(k)" is simply the application of an advanced understanding of the laws of the universe: a science.

The handful of topics covered in this chapter are by no means comprehensive. To cover *all* topics that could be considered "magic(k)" would require (at *least*) an entire volume. One day, I may write that volume; but

---

[†] If you have not yet read any of these; or if those that you *have* read simply failed to convince you, then I highly recommend Dean Radin's recent work "Supernormal."

[††] For example: if your mind says "I want that cup over on the table," and your mind causes your body to get up, walk to the table, and grab the cup; that is *not* "magic(k)"—your mind achieved its goal by means of standard "physical world" cause and effect. "Magic(k)" would be thinking "I want that cup," and having the cup either float or teleport into your hand.

in the meantime, this chapter will at least provide the reader with a framework that will allow them to evaluate and explain any phenomenon that might be commonly considered to be "magic(k)."

The holographic sub-quantum mind is embedded within the sub-quantum infrastructure of the universe. Thus, our minds are capable of *directly* interacting with the sub-quantum vortices and pixel grid—the most fundamental basis of what we call "matter"—as our minds are part of the same infrastructure. This is the basis for such phenomena as "psychokinesis," wherein the mind produces a direct effect upon an object within the "physical world" without any intervening physical means. This is why it has been observed for thousands of years in various cultures that form always follows energy. Chinese medicine and acupuncture are based upon this very principle.

Imagine if you were to make a snowball and roll it down the side of a mountain: by the time that snowball reached the *bottom* of the mountain, it would be *much* larger, moving *very* fast; and would destroy whatever it hit. Similarly, very minor changes at the most fundamental levels of the sub-quantum infrastructure can produce significant effects by the time they percolate up to the "physical world."

The fundamental basis of "magic(k)" is fully grasping the concept of "no matter, no motion, no time." Once you comprehend the illusory nature of the "physical world" and all of its aspects, you can open your mind to the possibility of what was referred to in previous chapters as "enabling cheat codes"—the ability to influence the illusions of the "physical world" with your mind.

Understanding this intellectually is one thing; but resonating with it on a deep, visceral level is another. That is why the three requirements for accomplishing "magic(k)" are: belief, desire, and expectancy. It is easy to desire something; more difficult to believe that it is *possible*; but the most difficult to accomplish is *expectancy*. Only once your limited "physical world" ego consciousness is in sufficient resonance with the truth that your holographic sub-quantum mind already knows—that the "physical world" and its attendant phenomena of matter, motion, and "linear time" are an illusion—will you truly *expect* your mind to be capable of altering aspects of the illusion. This is why a cumulative approach is best: one that begins with small successes, then builds upon them to achieve ever-greater effects upon the "physical world." This approach substantiates belief, and fosters expectancy.

The "Holographic Cloning" effect that we discussed in chapter seventeen clearly demonstrates that exceedingly subtle influences can alter what we perceive as "solid matter." The fact that the objects being influenced with holographic cloning happened to be living matter is irrelevant. Inevitably, technology will be developed to exploit this same effect for altering *inanimate* matter as well. Your mind can easily accomplish similar feats *without* technology. We already see this at work with the "placebo effect" in medicine, as well as its opposite, the "nocebo effect." Most people refer to victims of the nocebo effect as "hypochondriacs," who often psychosomatically make themselves ill, or cause an illness from which they are already suffering to become worse.

But once again, these effects both involve *living* systems. Until technology accomplishes comparable feats with inanimate objects, the "belief curve" will remain quite steep. Thus the need for the previously mentioned method of slowly building upon small successes, until the "belief curve" has been surmounted, and expectancy becomes as natural as breathing.

One example of this effect in action was the first runner to break the "4 minute mile" barrier. Within a few years of this record-breaking event, several other runners were also able to accomplish this feat; a feat that was previously believed to be impossible. One man's success at achieving the "impossible" shattered the "belief barrier" for other runners, who thus proceeded to push themselves harder now that they knew for a *fact* that the goal was achievable. A similar progression works equally well for gradually loosening the bonds imposed by accepting as real the illusory "physical world" aspects of matter, motion, and "linear time."

The equivalent of the "4 minute mile" for our purposes is the ground-breaking research conducted by such steadfast scientists as J. B. Rhine and, in more recent years, Dean Radin. What they have scientifically documented in the lab demonstrates that the mind absolutely *can* affect the "physical world." After you prove this to yourself by reviewing their work, you may proceed to conducting your *own* experiments, replicating their results. This will set your foot firmly upon the path to achieving feats of what has traditionally been referred to as "magic(k)."

Energy

Vast quantities of information have been written over the past several *thousand* years about Chi, Ki, Kundalini, Prana, Od, Orgone, etc. (whatever each culture throughout history has chosen to call the energy that most people today recognize—thanks to movies and widespread Tai Chi practice—as "Chi"), so there is no need to reiterate the basics regarding this human bio-energy. If you are not already familiar with this subject, there are any number of decent volumes that will help get you up-to-speed.

The only point that must be made here, is that irrespective of which name one chooses, this energy is synonymous with the sub-quantum vortices: Tesla's "dynamic ether." It is important to understand this, as much of the material in the remainder of this chapter will rely upon a comprehension of this basic fact.

The most critical aspect to understand about this energy, is that in the "physical world," form *follows* energy. This is why learning to mentally control your own bio-energy is so important. Not only can it benefit your health, since the body is also subject to the "form follows energy" rule;[†] but controlling this energy within your body is the first step toward mentally molding your external reality by influencing energy that exists *outside* of your body.

[†]This is why Qi Gong (moving Chi with the mind alone) is much more effective than acupuncture, as it is able to control the 85% of the Chi that flows *through* the meridians, while acupuncture needles can only influence the 15% of the energy that flows (in the opposite direction) along the *outside* of the meridians.

The "Power Of Three" in traditional "magic(k)" is rooted in the fact that the average human brain produces between 7-10 Watts of power (the maximum possible power output of a human brain is roughly 25 Watts); but 20+ Watts are required to produce changes within the "physical world." This is why the minimum requirement for a traditional coven is three people; and why a group of only two is called a "working couple," not a "coven." This is also why the term "bright" is used to refer to a person who is of exceptional intelligence (or why we say that someone of low intelligence is "dim"); and why a light bulb came to be used as the symbol for someone having an idea.[†]

By utilizing advanced breathing and meditation techniques, a person can increase the power output of their brain to above 20 Watts, and thereby become capable of producing changes in the "physical world" on their own. That said, doing so is by no means the most efficient way to produce physical effects with the mind alone; simply the epitome of the "energy manipulation" method.

I am often asked "Exactly *how* does the mind move energy?" In the same fundamental way that firmly striking a low "E" key on a piano makes the "E" strings on a nearby guitar vibrate: resonance. Our holographic sub-quantum minds are distributed throughout (i.e., embedded within) the infinite holographic infrastructure of the universe. Thus, our minds are in constant resonance with the most fundamental level of reality. And since everything contained within the illusory "physical world" (including our bodies and the energy that flows through them) is merely a holographic construct *created* by that fundamental infrastructure; you might envision it as our holographic sub-quantum mind playing "music" that causes the elements of the "physical world" to vibrate at its whim. This *begins* with learning to control the flow of one's own bio-energy; but with practice, leads to so much more.

---

† Such sayings find their way into the vernacular for a reason, just as saying someone is "feeling blue," or is "green with envy," or is "seeing red" when they are angry. All of these sayings originated due to the innate ability of every person to see (or otherwise sense) the colors of the human aura.

Manifesting

There are two basic *modus operandi* with regard to accomplishing what most people refer to as "magic(k)"—the "energy manipulation" method, and the "manifesting" method.

The "energy manipulation" method is still firmly mired within the "cause and effect" way of thinking. And within the illusory "physical world," cause and effect *is* absolutely valid; just as within the context of the video game discussed in chapter three, you must treat the walls as "real," as your character is unable to simply walk through them. You obviously realize that neither the walls nor your character are "real" in any objective "real world" sense; and yet, if you wish to play the game, you must abide by the rules. (Unless you enable "cheat codes.") The "energy manipulation" method of achieving "magic(k)" abides by the rules of the "physical world," fully obeying cause and effect. This is why, as useful as such techniques can be for personal health, healing others, and performing minor feats of what most would call "psychokinesis/telekinesis;" ultimately, energy manipulation is a very limited method with regard to "magic(k)." To achieve any significant results, we must look to the *other* method: "manifesting."

Manifesting is, referring back to our video game analogy, "enabling cheat codes"—realizing that, from the universe's standpoint, there is no difference between "empty space" and a brick of solid gold.

If we remember the discussion concerning "illusory density" within the Gravity chapter, we can see how this bears upon manifesting. The universal sub-quantum pixel grid does not "care" what it "displays." Whether we hold out an "empty" hand, or a hand that is (from our "physical world" perspective) holding a gold bar; it makes no difference to the pixel grid, as nothing really changes. The exact same number of sub-quantum "pixels" are involved *either way*. It is only from the perspective of our limited "physical world" ego consciousness that it *seems* to make a difference.

In the "Holographic Cloning" chapter, we discussed

how one living object can be transformed into another living object, by expressing the holographic sub-quantum signature of the source subject onto the target. This capability can be extended, not only to transforming any *inanimate* object into another object; but to producing the desired object out of what most would call "thin air."

The immediate objection raised by most is that some sort of "blank" is required; as in the holographic cloning experiment. But as we discussed in the chapter on gravity—in the "Illusion Of Density" subsection—what we call "solid matter" is, in reality, no such thing: it is nothing but *empty space.* Thus, at the most fundamental level, "thin air" is really no different than any "solid object" we might wish to materialize within that "empty space." The apparent density of any "solid object" is as much an illusion as the "magic force" that keeps two strong magnets in repulsion mode apart. Mentally manipulating the elements of the sub-quantum pixel grid to "display" a so-called "solid object" where we previously perceived only "empty space" is most commonly known as "manifesting."

Traditionally, this particular form of "magic(k)" has been limited to such advanced fakirs as Sai Baba; or such rare western adepts as A. B Glaser.[†] But by understanding the fundamental underlying principles, science will eventually develop the technological capability to accomplish the same feat. The same means of storing and expressing a holographic sub-quantum signature onto a target organism or object as was detailed in the "Holographic Cloning" chapter could just as easily be utilized to impress that stored signature onto "empty space."

[†]When it comes to modern day feats of manifesting, it is to be noted that none of the debunkers such as "The Amazing Randy" and his ilk are able to replicate such feats under controlled conditions. Such stage illusionists as David Copperfield et al can work wonders when they are permitted to set-up their elaborate illusions in advance; and when the audience is not allowed to inspect the stage or its props. But they are unable to perform such feats under controlled conditions while being closely scrutinized and filmed. In sharp contrast, such modern adepts at manifesting as Sai Baba and A. B. Glaser have done so on many occasions. Sai Baba routinely produced up to 1,600 lbs. of "holy ash" during his demonstrations (often with film crews watching and recording his every move; not that he could possibly carry around 1,600 lbs. of hidden material in any case), while A. B. Glaser transmuted the fillings in people's teeth from standard dental amalgam into gold. Both of these men were keenly aware that the "physical world" is an illusion, and that the universe's infrastructure does not have a preference for how illusory holographic projections within the "physical world" are arranged.

Similarly, one might be able to simply record the holographic sub-quantum signature of "empty space," then utilize that recording to effectively "disintegrate" unwanted or dangerous objects, such as trash, nuclear waste, toxic medical or industrial waste, etc.

Extending this yet further, one could produce electricity by recording the holographic sub-quantum signature of a fully energized "pole pig" (the large transformers on electric poles that carry 14,400 volts to an entire block of houses, then step it down to 240 Volts at each house), and then impress that stored holographic sub-quantum signature onto a non-energized pole pig, thereby gaining the resulting 14,400 volts for free. (Minus, of course, the cost to power the device that expressed the holographic signature.) Essentially, we will have "energized" the pole pig to full capacity without the usual input from burning coal/gas, turning generators, etc.—we simply imposed the "charged state" onto a non-charged pole pig. This would provide free electricity via a feat of technological "manifesting."

This could create a truly "green" closed-loop system for any house or business. Water would be produced at the point of use, at exactly whatever temperature is required; and when it goes down the drain, "empty space" is projected onto the contents of the drain. The same with all toilets, garbage disposals; and for that matter, the garbage can. Clothes? No more washing. You simply "disintegrate" the clothes when you are done wearing them, and "project" a new set of whichever clothing you want for the next day.

If this all sounds like pure science fiction, you should be aware that the Stanford Linear Accelerator Center in California has already created matter from energy, by colliding photons using two extremely powerful lasers. It will only be a matter of time before they advance from using their current "brute force" approach, to a more finessed methodology based upon an understanding of the universe's sub-quantum infrastructure. It is appropriate that Wheeler was one of the first to propose—in the 1930s—that matter could be created from energy; and it was upon his work that the Stanford Linear Accelerator Center has based their modern research. If you remember our discussion of Wheeler from the first part of this book, he and David Bohm were two of the main proponents of a sub-quantum level below the Planck Length. As I said; it is only a matter of time.

But while *technological* manifesting may be realized at some point in the future, advanced adepts have been capable of such feats utilizing only the power of their *minds* for millennia. How can this be accomplished?

In addition to being able to control the flow of energy *within* the body, our mind is also capable of influencing energy *outside* of the body, and thereby affecting the "physical world." In a limited capacity, this is often known as "PK" or "TK"—psychokinesis or telekinesis. I say "limited," since using the mind to influence energy flow is the least effective method of mentally affecting the "physical world." It is, in effect, using one illusion—the energy—to influence other illusions: "physical objects." Your holographic sub-quantum mind, embedded as it is within the sub-quantum infrastructure of the universe, is fully capable of affecting the "physical world" *directly*, without need of energy as an intermediary.

The only reason we no longer see this ability being exercised more frequently (i.e., why we no longer see blatant acts of "magic(k)" on a regular basis as was reported in times gone by) is that our minds are temporarily immersed via "suspension of disbelief" within the illusion we perceive as the "physical world," in much the same way that you can become so involved in a book that the outside world fades away completely: your mind is temporarily immersed with the book's story. Similarly, from birth-to-death, most of us are focused on the *consensus reality* we call the "physical world" to such an extent that we are unable to directly access the holographic sub-quantum mind, and see this world for what it truly is. As the number of humans on the planet has increased, and the percentage of humanity that is "modern" and "civilized" has likewise increased, the *consensus reality* has been strengthened (via the "Maharishi Effect," to be discussed further on in this chapter) to the point that it has become exceedingly difficult for individuals to overcome the imposed constraints of said consensus reality.

This is why so few people today exhibit the ability to affect the "physical world" with their thoughts: the part of the mind that maintains "suspension of disbelief"[†] in order to keep the limited ego consciousness focused on the illusory "physical world" will not *allow* full access to the holographic sub-quantum mind. Whereas once it was easier to "cheat," and partially/temporarily disable the "suspension of disbelief" mechanism; today, the increased strength of the global *consensus reality* largely prevents that, thereby requiring an extremely focused mind in order to reach this state, and thus limiting the number of people capable of achieving such feats.

Once you are able to disengage "suspension of disbelief" at will, you will be capable of what most people call "magic(k)." But there exists a fine line between mastering such abilities, and insanity. Once "suspension of disbelief" is disengaged, many people find it difficult to *re*-engage this mechanism; and thus, they are unable to fully reintegrate into the consensus reality. This occurs quite frequently when psychoactive drugs are used to achieve such a state when the limited ego consciousness was not properly prepared for such a drastically different perspective on reality. This is why small steps are best, allowing you the opportunity to *slowly* learn to disengage and *re-engage* the "suspension of disbelief" mechanism *at will*. Think of it like watching *half* of a movie, leaving the house to run errands, then returning to the movie at some later time: picking-up right where you left off. *That* is the skill that is required to successfully disengage "suspension of disbelief" for the purpose of accomplishing "magic(k)," and later *re*-engaging "suspension of disbelief" in order to return to "consensus reality." This faculty is generally known as "compartmentalization." We see this

---

[†] "Suspension of disbelief" is what allows you to immerse yourself in a movie and enjoy it, rather than watching it critically and picking it apart as a film critic must do.

at work in a limited way with the previously discussed "Raikov Effect." But while Raikov's technique requires the assistance of a highly skilled hypnotist, thus limiting your results to the scope of said hypnotist's experiment and/or agenda; if you learn to engage this faculty on your own, the sky is the limit.

This helps us see that the ability to focus the mind is the true secret of "magic(k)." Remember, the universe itself is holographic in nature; and how are holograms produced? By a laser beam, as we discussed in chapter eight. But why a laser beam, and not a light bulb? A 40-watt laser can burn holes through metal, while a 40-watt incandescent bulb puts out very little useful heat. What is the difference? *Focus!* The light from a 40-watt incandescent bulb diffuses too rapidly to be useful, while the coherent light from a 40-watt laser delivers much more energy to the target. Similarly, if one wishes to effect changes within the holographic sub-quantum infrastructure of the universe, one must employ a *focused* mind, not the mental equivalent of a 40-watt light bulb. This is where breathing exercises and meditation come into play: they help you learn to focus your mind, and allow you to interact with the sub-quantum infrastructure within which the holographic sub-quantum mind is embedded.

This explains how altered states of consciousness, such as hypnosis and somnambulism, allow one to "bend the rules." The conscious mind—the mind that accepts the consensus reality—is temporarily "out of the way," allowing one's holographic sub-quantum mind to manifest effects in ways that *it* knows are possible; ways the conscious mind would never allow.

Teleportation

Manifesting lays the groundwork for teleportation. If you can cause so much as a *penny* to materialize within your "empty" hand; then you can mentally teleport yourself to any location in the infinite universe just as easily. The secret is learning to think of it, not as "moving" from one place to another; but to cease being "displayed" at one location, then resume being "displayed" at some *other* location.[†] Since information travels instantaneously through the infinite sub-quantum infrastructure of the universe, no hypothetical "wormhole" is necessary for teleportation. All that is required is for the holographic sub-quantum mind to alter its focus regarding where your "physical body" is "displayed." Since distance is as much an illusion as matter, motion, and "linear time," it really makes no difference how the illusion appears to be configured; whether that means materializing a penny in your previously "empty" hand; or "teleporting" your "physical body" trillions of light years away.

Once a technological infrastructure capable of interacting with the sub-quantum infrastructure of the universe has been developed, both general manifesting as well as the specialized case known as teleportation will be accomplished via technology, bringing to the masses what has traditionally been the arena of the "spiritual adept."

A Brief Note On The Immutable Nature Of Space

Most people are familiar with the old Physics adage "matter can neither be created nor destroyed." And yet, we regularly hear talk within the scientific mainstream of "warping space." Nothing of the kind can happen; neither via technology, nor even mentally. The "physical world" is an illusion; and like a movie being displayed on a TV screen, the content of the movie can in *no way* affect the pixels of the television.[††] Similarly, the

---

[†] While we are used to the illusion of "linear motion" (which fits hand-in-glove with the illusion of "linear time") here in the illusory "physical world;" in truth, there is no need to travel through all of the usual points between "point A" and "point B." Much as dragging your pointer around your computer screen is the norm; but with the trackpads on some laptops, you can enable a feature wherein pressing with your thumb at one corner of the trackpad, then releasing quickly as you then press the opposite corner with a finger will "teleport" your pointer icon to the opposite corner of the screen without having to travel through all of the points in-between. (Often useful with today's exceedingly large computer screens.) In the same way that the computer is capable of ceasing to display the pointer icon at one point, and immediately displaying it at any *other* point on your computer screen; your holographic sub-quantum mind is fully capable of ceasing to "display" your so-called "physical body" at your present location, and displaying it instead at any other location of your choosing. This may not be the standard "consensus reality" *modus operandi* for movement within the "physical world"—it is definitely "cheating;" but that is, after all, what "magic(k)" is all about.

[††] And before some weisenheimer AV Tech mentions "burn-in"—yes, I am completely aware of this phenomenon. This is, after all, an *analogy*. So assume I mean that the content cannot affect the pixels in any *immediate* fashion.

sub-quantum pixel grid cannot be affected by holographic illusions within the "physical world." Yes, we can "enable cheat codes," and alter the *illusions* that are displayed—such as with manifestation/transmutation of objects, telekinesis, teleportation, etc.—but not the fundamental sub-quantum infrastructure of the universe *itself*. So while technology will eventually be capable of teleporting matter across vast distances within the "physical world," it will not be accomplished via the oft postulated "wormhole;" but rather, by simply ceasing to holographically project an object at one location, and resume projecting it at another; similar to the "trackpad/ pointer icon" trick discussed previously.

On The Purpose Of Ritual

Most scientifically-minded people dismiss ritual as a mere remnant of ancient, backward belief systems; and in the vast majority of cases, that assessment is spot-on. But if utilized properly, ritual *can* serve a purpose during the beginning phases of learning what most people refer to as "magic(k)."

It is one thing to think "I am hungry; I should really get up and make some food." But that is simply a desire. Only once you set yourself in motion to actually *make* the food are results obtained. That is how our minds are accustomed to events occurring in our day-to-day lives. Thus, when we want to "manifest" specific objects or outcomes—at least in the beginning phases of learning this ability—we must give the mind what it has come to expect: some form of physical impetus to "set the wheels in motion." Hence, the power and purpose of ritual.

Gods, Demons, Elementals, And Other Thought Forms

Manifesting is what allowed ancient people's devout "belief" to create a "god," "demon," "elemental," or other "nature spirit." (What the ancient Tibetans called a "thought form," or "Tulpa.") What they had created via their alternative consensus reality is a group holographic sub-quantum mind that eventually took on an individual consciousness of its own. To what extent could such a creation sustain existence on its own, once the "believers" faded away? Can such a group-consciousness be sustained, once created, if the individual minds that created it withdraw? An intriguing thought to ponder for those interested in the history of pagan summoning rituals and practices.

But not all such entities were created by "believers." In ancient times, many of the "gods," "demons," "avatars," etc., were the equivalent of today's "hackers"—people who did not want to play by the rules of the game. And since "consensus reality" was still relatively weak and malleable due to the relatively small number of people on the planet, the "hackers" were able to wreak havoc for a period of time. But as the global population grew, and became more interconnected and "civilized," the consensus reality gained strength as the accepted paradigm; which is why only the Catholics and a few remote shamanic societies still experience "miracles" in this day and age where the consensus reality reigns supreme.

But sub-quantum physics may be the fly in the ointment of the established consensus reality. At long last, the cold, hard light of scientific logic may be showing us that there *is* more to see beyond the veil. A *new* consensus reality is gaining strength, and it will only be a matter of time before it is commonplace for people to "peer beyond the veil."

The Maharishi Effect

The "Maharishi Effect" is the documented ability of a group of individual minds focusing together to affect the behavior of a greater number of minds in the vicinity. The effect requires a specific ratio of "focusing minds" to "local population" minds in order to become apparent. i.e., The "focusing minds" must exert sufficient influence to overcome—even if only in relatively *minor* ways—the "consensus reality" normally adhered to by the majority of the local population.

Form follows energy, allowing thoughts to manifest

as "solid objects" or "physical world" situations, since energy—itself an illusion of the "physical world"—is "displayed" by the universal sub-quantum infrastructure. If enough people focus their minds upon specific mental constructs, these objects and/or situations can take on an apparent permanence, independent of any individual. Many people who have studied astral projection are familiar with this effect within the "astral world,"† but fail to realize that the same effect also takes place within the "physical world"—it simply requires more time, and considerably more mental focus. And a thorough understanding of the underlying sub-quantum physics certainly helps.

You cannot engineer a system to reliably produce desired effects until you fully comprehend the fundamental infrastructure upon which said system will be built. Case in point: the scientific mainstream does not yet understand the fundamentals of gravity—they can measure *what* it does, but have no understanding of *why*. Thus, they still strap our astronauts to large sticks of dynamite in order to send them into space, occasionally with tragic results. Once they understand the true fundamentals of gravity, they will be able to send people into space much more safely and efficiently.

In much the same way, while millennia of trial-and-error have allowed certain people—and even certain *groups* of people—to produce paranormal effects on a somewhat repeatable basis; in order to produce such effects more efficiently and reliably, a thorough understanding of the holographic sub-quantum mind is required. Understand the underlying infrastructure, and you can more easily engineer systems to produce the desired results.

Imagine someone that has never seen a wheel, much less a car (perhaps someone plucked directly out of the

---

† With *billions* of human minds presently lending energy to their various belief systems (not to mention the energy given to those constructs by generations *past*), it should come as no surprise that people find what they *expect* to find while traveling the "astral planes" in the "out of body" state. This is why Christians who find themselves "out of body" in the astral levels (as opposed to hovering over their body in an operating room, for instance) may see "Jesus", or other appropriate religious figures, while Buddhists might see Buddha, etc.

stone age); and then we ask them to *move* a car. They would have to work very hard to shove that car even a few feet while it is still in "Park." Since we understand how cars work, we would be able to accomplish the same task much more efficiently than our stone age friend. With modern rocketry—as well as modern psychology and parapsychology—the scientific mainstream is still trying to push the car while it is in "Park." Understanding that the universe is holographic in nature is like realizing that the car has wheels, and could potentially be made to roll. Grasping that the mind operates on a sub-quantum level provides the keys with which to *start* the car. Learning to actively interact with the sub-quantum infrastructure of the universe in order to control the illusory "physical world" allows us to put the car in gear and drive away.

Revisiting for a moment the subject of astral projection—especially the works of Robert Monroe—the similarity between the "physical world" and the "mini-worlds" created by all of us while we sleep becomes readily apparent. It was observed by Monroe that, the greater the number of people who "believed" in a specific other-worldly construct, the more "real" it became, even for other visitors who were not involved with its creation. (Imagine, for instance, the apparent "reality" that Purgatory, Heaven, and Hell must enjoy; not to mention the Tibetan "Bardos.")

We can easily extrapolate that what we call the "physical world" may simply be another holographic construct, created by countless individual minds over time via the same "consensus reality" principle; similar to the way in which DNA has evolved:[†] by constant reinforcement of its holographic sub-quantum signature,

---

[†] DNA is an emergent phenomenon, another example of fractal expression. Simply put: form follows energy; and energy is comprised of the sub-quantum vortices. The same underlying infrastructure responsible for gravity, current flow, magnetism, and all of the other phenomena of our "physical world," is also responsible for the holographic data storage system that is DNA. Most people think of DNA as a blueprint. It is obviously a blueprint *now*; but this was not always the case. Consider the analogy of roads. In modern times, some highways are designed by city planners, and put in a specific place for a specific reason, often where no road has ever existed before. Other roads began as foot-worn paths that slowly became more defined with increased use, and eventually evolved into paved highways. Rather than being a "pre-planned highway," DNA is actually a "well-worn trail" that eventually evolved into the blueprint we see today.

until the "created" seems to be (and, in our "physical world," actually *acts* as) a blueprint in its own right.

Imagine that an author writes a science-fiction novel wherein they posit a physically infinite universe that has always existed, and can never end. We must accept that premise if we are to "suspend disbelief" long enough to enjoy the story that is set against that background. What we call the "physical world" may have arisen in a similar fashion. As we saw in the "Many Worlds" chapter, all possibilities are already encoded within the infinite sub-quantum infrastructure of the universe. An infinite (with respect to "space" as well as "time") "physical world" *can* be created, simply because it can be *posited*. If, indeed, the "physical world" we believe we inhabit is holographic in nature, and is simply another "whirlpool" within the sub-quantum "river;" then there would be no logical objection to creating such an infinite construct where none existed before (from the perspective of the rest of the infinite spectrum of consciousness across which our holographic sub-quantum minds are active); in much the same way that, within the realm of imagination, no-one could forbid an author from positing any setting they wished for their novel.

Thus, the entire infinite "physical world" could have originally resulted from nothing more than an act of manifesting—a collaborative effort between various holographic sub-quantum minds. And while holographic sub-quantum minds can *create* such a "physical world," they can also m*odify* it. This is seen, not only with manifesting, but in the "Maharishi Effect."

"Miracles" Ancient And Modern

In ancient times, when people were less sophisticated and "civilized," and when there were far fewer of them, reality was much less "fixed"—there was much less of a global "consensus" than exists today. This is why "miracles," "magic(k)," "gods," "demons," "nature spirits," and other such phenomena were much more prevalent in ancient times, and are so rare today. Modern civilization's "consensus reality" prohibits, for the most part, such "miracles" from occurring.

Within the *overall* global "consensus reality," there are, to varying degrees, "consensus sub-realities." Perhaps the largest of these—or at least the most well-known and organized—is the Catholic Church. This is why we still see so-called "blood miracles," such as bleeding statues, stigmata, and transubstantiation within the Catholic Church: the "consensus sub-reality" of the Catholic Church—a *group* holographic sub-quantum mind, as discussed in chapter fourteen—accepts these phenomena, and their occurrences are positively reinforced as being of religious significance. Thus, Catholics continue, even in the modern era, to experience these "blood miracles." It is strictly their "consensus sub-reality" that allows this to happen.

We discussed in the last section how the "hackers" in ancient times were able to "cheat" the system; at least, until the global "consensus reality" became too strong. It was easier to break out of the accepted "consensus reality" back then, because it was not as well reinforced as it is today—there were far fewer people on the planet. But over the millennia, as the global population has increased, "consensus reality" has grown considerably stronger.

But today, the global consensus has begun trending the other way; largely because the *old* "consensus reality" was mediated by a global elite who fed the public well-regulated information, suppressing whatever did not fit the paradigm they wished the masses to accept. With the advent of the Internet, the ability of people who experience events that do not quite fit within "consensus reality" to communicate with one another has improved exponentially. The average person can now easily amass a wealth of data on occurrences that do not fit neatly within the bounds of the accepted "consensus reality." And the more such occurrences become "common knowledge," the more it weakens the "consensus reality," thereby facilitating the increased incidence of such occurrences. You can see the self-reinforcing cycle that has begun.

A revolution in consciousness is upon us, and the world will become an increasingly interesting place as it unfolds. The years to come will be anything but boring.

The Myth Of Karma

At this point in my lectures and seminars, many people begin to worry about the effect that exercising such abilities might have upon their "karma." The gist of their logic usually goes something like this: "If I can incur Karmic Debt during my lifetime *without* having any special abilities, wouldn't using such abilities—like the Maharishi Effect that can affect large *groups* of people—open me up to even *more* Karmic Debt?" Fortunately, we need not worry in this regard, as "karma" is a complete fiction.

Traditional notions of "karma" teach that you "get what you deserve." In actuality, there is no such thing as "karma"—you do not get what you "deserve;" you get what you *expect*. Your beliefs and expectations shape your world. Thoughts manifest as "solid objects" or "physical world" situations. Thus, you can bring your *own* "karma" upon yourself if you feel guilty about actions you have taken, or feel that you "deserve" to be punished.

And while traditionally, "karma" is supposed to affect your *next* life, this effect is often evident during a single lifetime, as it depends solely upon the ability of the individual to manifest their *own* "karma," in much the same way that a hypochondriac manifests their own illnesses. Our thoughts shape the sub-quantum aspects of ourselves; aspects which do, eventually, "trickle up" to our "physical world" situation; like the snowball rolling down the snow-covered mountain, gaining size and momentum as it goes. Your thoughts can help you, or they can harm you. This is another reason why learning to meditate—to observe and control your own thoughts—is so important.

To assist you with feelings of guilt, remember that the "infinite 'Choose Your Own Adventure' book" that comprises our holographic sub-quantum pixel-grid-based "reality" is, by definition, different for each one of us. This means that, not only are we free to fully create the world around us as we see fit, but nothing we do really matters. I do not mean this in a *Sartre-esque* "nothing really matters" way; rather the opposite: that we are *truly free* to do *anything* we wish, because we *cannot* possibly harm anyone else.[†]

Not only can other people not be *truly* harmed, as their holographic sub-quantum mind is *permanent*—etched into the very infrastructure of the infinite holographic universe—but even any temporary/situational *perceived* "harm" is illusory, experienced only in *our* version of the universe; one particular path through the infinite "Choose Your Own Adventure" story. We must always bear in mind the ancient Hindu adage: "It's Maya."

Sympathetic Magic(k)

"Sympathetic magic(k)" covers a wide range of topics, from witchcraft, voodoo, and psionics, to such "remote healing" modalities as Reiki. The fundamental aspect common to all forms of sympathetic magic(k) is the establishment of a holographic sub-quantum resonance between the subjects. In witchcraft and voodoo, one generally sees an item belonging to the target being used to create this link, whether the purpose is to harm, charm, or to heal. Just as in psychometry, the personal item belonging to the target has been imprinted with their holographic sub-quantum signature, allowing the practitioner to establish a resonant connection to the target. But where sympathetic magic(k) differs from psychometry is that, while psychometry is only used to gather information, sympathetic magic(k) seeks to *influence* the target.[††]

[†]Though certain actions can lead to very real negative consequences in the current storyline you are following in the "here and now;" just as making poor choices in a video game can get your character killed. Make your choices accordingly, but spare yourself the guilt by remaining mindful of the larger picture.

[††] Not to split hairs; but in reality, even when seeking to remotely influence a subject—whether for constructive *or* destructive purposes—all that is being exchanged is *information*. The difference lies in the *intent* of that information exchange. With remote influencing, the purpose is to utilize the information exchange to initiate physical changes within the target; more specifically, to induce *them* to produce said physical changes (i.e., manifesting), either within themselves, within those near to them, or within their immediate physical environment. Think of it as the difference between two people having a pleasant two-way conversation with no particular agenda, as opposed to one person using psychological techniques to actively manipulate the other party in the conversation. But in the end, it is still only information that is being exchanged.

Reiki and similar "remote healing" techniques differ from witchcraft and voodoo in that an object connected to the target is not necessarily required: the practitioner is generally able to establish a resonance with the target in much the same way as a medium.

Of all the methods of sympathetic magic(k), psionics is the most varied. Some psionic "devices" require an object in order to establish a resonance with the target, while others do not. But in all cases, psionic "devices" only act to *focus the mind*. The devices themselves are actually not intrinsically important; similar to talismans in traditional witchcraft.[†]

Anything that is accomplished via the use of a psionic device, can also be accomplished *without* the use of such a device, once the person learns to focus their mind without need of an external aid. Think of psionic devices like the training wheels on a child's first bicycle.

Sacred Geometry

Sacred geometry effects are produced via a combination of sub-quantum resonance and fractal expression. This is why we see the "golden ratio," not only within DNA and *living* creatures, but also in the patterns produced by plasma and electrical discharge, the behavior of gravity, and the shapes of galaxies. And just as in harmonic resonance with musical notes, the closer the harmonic, the greater the coupling. Thus, "sacred geometry" exerts a *subtle* influence that taps only the most minuscule amount of *power*; but this is enough to produce profound effects upon the human psyche and bio-energy system. (More on that in the "Astrology" chapter.)

In Summation

In the "Many Worlds" chapter, we discussed the sub-quantum pixel grid as being akin to a dictionary of events, interactions, and experiences from which any

---

[†] Similarly, with many of the other traditional tools and techniques of the occult— pendulums, Ouija boards, automatic writing, diving rods, magic(k) mirrors, crystal balls, etc.—the devices and techniques themselves are *not* what is important: they are simply tools that aid in focusing the mind.

"story" could be constructed. But in the "Choose Your Own Adventure" book analogy from which this example flows, you are merely making choices between pre-defined paths that have been written by someone *else*. With "magic(k)," *you* become the author, writing your *own* story, rather than being confined by the limited choices presented to you by someone else. If you wish to attain the highest degree of true freedom, you must write your *own* story.

Just as in life, you can enjoy reading books that other people have written; but you can also choose to write your own.[†] This is how all of the "avatars" noted throughout history have been able to accomplish their amazing feats: they chose to write their *own* stories.

Since the sub-quantum pixel grid is an infinite dictionary of events, interactions, and experiences, you are free to draw upon that unlimited database for the purpose of assembling whichever "stories" you wish. The vast majority of people choose to be *readers* rather than authors; and that is fine: that is the path they have chosen, the experience they wish to have. But for those who find themselves yearning for more; take the reins, and write your *own* story. *That* is the ultimate expression of free will; and it is also the secret to "magic(k)."

---

†This analogy is just as applicable to movies or video games: you can watch movies written and produced by others; or you can write and produce your own. You can play video games created by others, or you can learn to create your own.

Chapter 23:
Astrology

*"That we can now think of no mechanism
for astrology is relevant but unconvincing.
No mechanism was known, for example, for
continental drift when it was proposed by Wegener.
Nevertheless, we see that Wegener was right, and
those who objected on the grounds of unavailable
mechanism were wrong."*
— *Carl Sagan*

First and foremost with regard to astrology, it should be noted that the constellations are not "real." This is immediately evident given that various cultures had different constellations. People simply observed what effect being born at certain times had upon a person, then "connected the dots" in the sky to construct a glyph that related to the observed personality traits. In other words, the observed mannerisms dictated the shape of the constellations, not the other way around.[†]

In the "energy" section of the "Magic(k)" chapter, I mentioned "Orgone," which was the name Wilhelm Reich gave to the energy that Tesla called the "ether;" what we know today as the sub-quantum vortices. One of the observations Reich made regarding Orgone was that wherever you find electricity and magnetism, you will find Orgone. Since we know from chapters five and six that what we call magnetism and electricity are nothing but higher-level expressions of the sub-quantum vortices, Reich's observation makes perfect sense—it is a given that wherever you find electricity and magnetism, you will find what Reich called Orgone.

Even when we consider the matter simply from the standpoint of standard plasma physics—as Wallace Thornhill and David Talbot demonstrated in "The Electric Universe"—you can see that, at the very *least*, the entire solar system (possibly the entire galaxy) is an interconnected electrical circuit. And as has been noted in psychological research going back many decades, a multitude of external stimuli—from scents, sounds, and sights, to tactile sensations—all have an impact upon our psyche.[††]

[†] Consider the example of cloud gazing: several different people can look at the same cloud, and each of them will see a different image depicted in that cloud. Similarly, any image one wishes can be constructed from the multitude of stars in the sky. It is much like the psychological tool generally known as the "Ink Blot Test," or the "Rorschach Test."

[††] That is why prison walls are painted with greens and blues—"calming colors." Research showed that if the walls were painted red, the prisons experienced a greater number of violent incidents, with "calming colors" having the reverse effect. It has been well quantified by modern psychology how colors affect mood and behavior. Thus, if you shine a white light onto someone, it will produce a specific effect within that person. Place a colored *gel* over the light, and it will produce a *different* effect. This is how planetary bodies in various configurations produce different net effects upon us. Astrological analysis is like knowing which colored gels were placed in front of a white light, to yield a colored light that affected your mood and behavior in a specific way.

Similarly, electric and magnetic fields also affect the human psyche.[†]

In his book, "Seven Experiments That Could Change the World," Rupert Sheldrake discussed the fact that, if you take a carrier pigeon that can find its way to its coop even if said coop is on a moving truck (in World War II the U.S. Military utilized this ability for sending messages; and the pigeons were able to find their coop no matter how far away it was moved), and you tape a tiny rare earth magnet to the back of the pigeon's neck; it will fly around aimlessly in circles, and never find its coop. That tiny magnetic field has a rather large effect upon the pigeon. So imagine all of the electromagnetic fields that are being emitted by the Sun and all of the planets in our solar system—each of these bodies having a different density, a different mineral composition, different atmospheric gases, etc. (meaning that their electric and magnetic characteristics vary widely)— bombarding the Earth from all directions in a multitude of ever-changing combinations based upon the positions and interactions of those bodies. Thus, depending upon how these bodies are arranged with respect to each other, the net effect upon people will vary.

And that is simply the *electromagnetic* aspect of the solar system's various planetary bodies. We must factor-in the fact that, along with the effects of the higher-level *expressions* of the sub-quantum vortices that we call electricity and magnetism, we also have the *direct* effects of the sub-quantum vortices themselves to consider. And since our *minds* exist at a sub-quantum level, it is self-evident that the sub-quantum effect of the various planetary bodies in the solar system upon our minds will be greater than the electric and magnetic effects. The analysis of these effects is what is commonly known as "astrology."

So why is the exact moment and location of your birth

[†] Electromagnetic and electrostatic fields affect our mood and behavior. Some modern workplaces and factories utilize negative ion generators for this purpose. Positive ion fronts that precede storms induce lethargy and muddled thinking, while negative ion fronts that *follow* storms have the reverse effect. Employers find that negative ion generators improve the mood and productivity of employees, thereby reducing errors and accidents.

important? Would not these external electric, magnetic, and sub-quantum influences affect you while you are still inside the womb? In fact, they do not; because the human body acts as a dielectric—meaning all frequencies that impinge upon your body are averaged-out over the entire *surface* of the body. So the mother's body acts as a shield against external influences while the fetus remains within. But as soon as the fetus emerges from the womb, and is directly exposed for the first time to the solar system's multitude of external electric, magnetic, and sub-quantum influences, the body is "imprinted" in much the same way that a paper clip exposed to a strong magnetic field will become slightly magnetic.[†] That is why the exact moment and location of your birth is important: because that is the moment that you are "imprinted" by the ambient external influences; an imprint that will not only affect your core personality, but how future external electric, magnetic, and sub-quantum influences will affect you. Think of it like a compass needle: once it is magnetized, one end will invariably point to the external influence we call the "magnetic north pole" of Earth. Similarly, once imprinted, your body will react in specific ways to future combinations of external influences; and since the body and mind function as a unit, this will affect your behavior. It is as unavoidable as a compass needle pointing north. But whereas a compass needle cannot do else but succumb to the external influence; we, as thinking beings, can acknowledge the influence, then utilize our minds to *overcome* said influence. But recognition of the *existence* of such external influences (as well as their quantification) is the necessary first step in overcoming them.

---

†Dr. Karl Ludwig von Reichenbach (discoverer of Eupione, Paraffin, Pittacal, and Phenol), similar to Wilhelm Reich in later years, conducted scientific research that led to the discovery of a more fundamental energy underlying electricity and magnetism. Reich called it "Orgone," while Reichenbach called it "Od." Reichenbach's pertinent contribution to the subject of astrology was the use of a telescope to "imprint" pure water samples with the energy signature of the visible planetary bodies in our solar system, including the Sun. Double-blind testing of these imprinted water samples showed a specific effect upon the behavior of anyone who ingested the water. Dr. Buryl Payne has recently replicated Reichenbach's water imprinting experiments. It is interesting to note the similarity to Dr. Masaru Emoto's experiments involving the imprinting of water samples.

Chapter 24:
Ghosts & Hauntings

"For who can wonder that man should feel a vague belief in tales of disembodied spirits wandering through those places which they once dearly affected, when he himself, scarcely less separated from his old world than they, is for ever lingering upon past emotions and bygone times, and hovering, the ghost of his former self, about the places and people that warmed his heart of old?"
— Charles Dickens

With regard to the subject of ghosts and "hauntings" in general; I wish to stress that, as with the previous chapter on astrology, this chapter does not seek to "prove" that ghosts or "hauntings" exist, as many volumes have been written to explore that issue in detail. Rather, we will proceed from the assumption that such phenomena *do* occur from time to time, and discuss the mechanism responsible.

Ghosts, or "hauntings," are nothing more than the result of a residual holographic sub-quantum imprint. In much the same manner as sound waves impacting upon a surface create sympathetic vibrations in that surface (with said vibrations being retrievable for up to an hour afterward via the previously discussed CIA technology), holographic sub-quantum signatures are similarly impressed upon their immediate environment. And whereas sound impressions are made rapidly, and fade just as rapidly; holographic sub-quantum signatures require considerably more time to create a lasting imprint within most substances, but are also much longer-lasting. Just as water cannot erode a shoreline overnight; but give it a hundred years, and you will see significant, lasting results.

Another example is a path through the woods: the first time you walk through, the foliage will spring back relatively quickly, leaving little trace of your passage. But if that route is traveled frequently, a path becomes worn, which later may become a road; and eventually even a paved highway. Similarly, when a person's holographic sub-quantum signature is impressed upon objects over a protracted period of time, a psychometrist or medium can utilize that imprint to "tune-in" to the holographic sub-quantum mind of that person.

If the impression is sufficiently strong, the impressed object (usually a building in which the person spent a significant amount of time) may act as a substitute for the body of a deceased person.[†] The degree to which this effect occurs obviously depends upon the extent of the imprint;

---

† It should be noted that, while not *common*, it is possible for even the smallest object to be impressed strongly enough with a person's holographic sub-quantum signature to produce the same effect.

and in no case would the resonance with the holographic sub-quantum mind of the deceased person ever approach the level of resonance with the deceased person's body. But the effect can be strong enough to create what is commonly known as a ghost, or a "haunting."

Thus, when a person lives in a building for many years, they leave a residual holographic sub-quantum imprint; so while their "physical body" may be gone, their holographic sub-quantum mind is still able to resonate with that building.

Often, when what psychiatrists call a "significant emotional event" occurs in a building, the emotional energy of that event can imprint a person's holographic sub-quantum signature into that structure much more rapidly. This can produce a stronger imprint than somebody simply dwelling there quietly, uneventfully, for many years. Thus, a holographic sub-quantum imprint *can* be produced over a short period of time; but as time length decreases, intensity must *increase*: they are inversely proportional.

If you have high-intensity, emotionally-charged events occurring at a location over a long period of time, that will obviously imprint the strongest possible holographic sub-quantum signature.

A "poltergeist" is a unique form of "haunting" phenomenon, as it requires a living person, still in their "physical body," providing high levels of energy to power the various associated phenomena. (As discussed in "Chapter 22: Magic(k).") This provided energy allows a strong holographic sub-quantum imprint within a building to *channel* that energy, similar to biasing a transistor. Or picture it as a hologram, with the energy of the living person being used as the laser beam that expresses the holographic image that is recorded into a piece of holographic film: nothing but "squiggles," until the appropriate frequency of coherent light is impressed upon the film.

The living person powering the poltergeist phenomenon is usually a prepubescent/adolescent, as they exude large quantities of chaotic, unfocused mental energy. Thus, rather than overwriting the existing imprint, they begin to *resonate* with the holographic sub-quantum mind of the deceased person who *created* the imprint. This is similar to a medium who allows the holographic sub-quantum mind with whom they are resonating to gain control; a state known as possession. With the poltergeist phenomenon, a living person who is in a susceptible mental state resonates with the holographic sub-quantum mind of the deceased person who created the imprint within the house; to the extent that the living person becomes possessed, thereby allowing the holographic sub-quantum mind of the deceased person to utilize the living person's energy to produce tangible effects within the "physical world."

I am often asked if it is possible to remove holographic sub-quantum imprints. Indeed it is. Such imprints are actually quite delicate, which is why many years are generally required to create them. A high-intensity *new* imprint can easily override the existing imprint. It is similar to accidentally putting your credit card near a strong magnet: the data is erased, as the magnetic field of the magnet is stronger than the imprint on the magnetic stripe of your credit card. It is relatively easy to "erase" the existing holographic sub-quantum imprint of a building or object by utilizing a powerful, homogeneous energy signature. This can be accomplished via technology— a mild EMP, running a Tesla Coil around the clock for a few days, or even protracted high-decibel "white noise" (preferably ELF, or infrasound, for maximum effect)— or by someone experienced with energy manipulation. All shamanic cultures have developed rituals for this purpose; though the power lies not in the ritual itself, but within the ability of the practitioner to manipulate energy fields.

This last point is key. Take, for instance, the Native American ritual of "smudging" to eradicate "negative energies" from a building. Anyone can go to the local herbal supply shop, buy some sage, and attempt to "smudge" their home. Generally, they will not obtain the desired result. The "magic" is not within the sage herb itself; it is in the ability of the *trained* individual who utilizes sage to aid them in achieving a specific state of *mind*, which then allows them to manipulate energy fields for the purpose of overriding the building's current holographic sub-quantum imprint.

The bottom line: all you must do is to modulate—by any available means—the energy signature of a location to a sufficient degree, and you will eradicate the previous holographic sub-quantum imprint.

It should be noted that, if a person with a strong, stable bio-energy field takes up residence within a "haunted" building; over time, that person's high energy level will also eradicate the previous holographic sub-quantum imprint, much like recording over a video tape. So while some type of technological or shamanic intervention is required if you wish to "cleanse" a place *quickly*; it can also be accomplished over time, with the rate depending upon the energy level of the new occupant.

Afterword

I truly enjoyed researching and writing this book. Though doing so has consumed the past fifteen years of my life, I feel it was time well-spent. And while I understand that most readers chose this book for the material contained within its *second* half; for me, the research required to produce the *first* half was the most rewarding. Picking-up where such great minds as David Bohm, John Wheeler, and Nikola Tesla left off, and thereby laying the groundwork for the creation of an entirely new technological infrastructure capable of directly interacting with the sub-quantum level of the universe; that has been a thrilling adventure, every step of the way.

The addition of the sub-quantum to the lexicon of Physics is akin to the addition of the "z" axis to two-dimensional geometry; and the results that will be produced by the hard-working experimental physicists and engineers who actually undertake the creation of the above-mentioned sub-quantum technological infrastructure will be equally as revolutionary as the addition of a third dimension to geometry.

I have already received comments stating that there are many additional topics that *could* have been covered in the second half of the book; and if I receive enough requests to address specific topics, I may either cover them in a future "expanded edition," or possibly— if the number of requested topics justifies it— in a sequel. I have already considered writing dedicated single-topic volumes, such as "Sub-Quantum Astrology" and "Sub-Quantum Healing."

In the meantime, I hope you enjoyed reading this book as much as I enjoyed writing it. My hope is that, rather than eliciting a simple "Well, that was interesting . . ." reaction, after which the reader simply moves on; it will instead inspire tangible advances in technology, as well as in the study of consciousness.

Many thanks,
Louis Malklaka